Water Power Development

by Adolph Black

with an introduction by Roger Chambers

This work contains material that was originally published in 1908.

This publication is within the Public Domain.

*This edition is reprinted for educational purposes
and in accordance with all applicable Federal Laws.*

Introduction Copyright 2017 by Roger Chambers

Self Reliance Books

Get more historic titles on animal and stock breeding, gardening and old fashioned skills by visiting us at:

http://selfreliancebooks.blogspot.com/

Introduction

I am pleased to present yet another title on Homesteading and Farm Life.

The work is in the Public Domain and is re-printed here in accordance with Federal Laws.

As with all reprinted books of this age that are intended to perfectly reproduce the original edition, considerable pains and effort had to be undertaken to correct fading and sometimes outright damage to existing proofs of this title. At times, this task is quite monumental, requiring an almost total "rebuilding" of some pages from digital proofs of multiple copies. Despite this, imperfections still sometimes exist in the final proof and may detract from the visual appearance of the text.

I hope you enjoy reading this book as much as I enjoyed making it available to readers again.

Roger Chambers

WATER-POWER DEVELOPMENT

PART I

1. Introduction. One of the fundamental teachings of science is that all energy in the solar system is derived from the sun. Through the agency of that luminary, water from the earth's oceans, seas, and lakes is transformed into vapor, and in this condition is diffused throughout the atmosphere, transported by the winds—themselves created by this same solar energy—over long distances and wide areas, and finally precipitated over land and water, hills and valleys, mainly in the form of rain and snow. Of the total precipitation on the continents, part is evaporated from land and water surfaces, vegetation, etc.; part runs off more or less rapidly as surface flow into the nearby drainage channels, and thence, more or less directly, to the ocean; and part sinks into the ground. Of this last, a portion is retained by capillary attraction within reach of vegetation, to be taken up slowly by the rootlets and transpired through the leaves; the balance percolates downward until it reaches the surface of the underground water flow, which it joins in its relatively slow motion to some nearby stream, lake, or other drainage course, or directly to the ocean. It is then again evaporated into the atmosphere, with a continuous repetition of the cycle described above.

Thus every elevated body of water, every running stream, is a source of power whose energy has been derived or borrowed from the sun; and under proper conditions, a large proportion of this energy may be transformed into useful work.

2. Unit of Work. For industrial purposes, the unit of work most generally adopted is the *foot-pound* (ft.-lb.), which represents the quantity of work done in lifting a mass of one pound through a height of one foot against the opposing force of gravity—or in raising a weight of one pound through a height of one foot. Since the force of gravity, and therefore the weight of a given mass, is not constant for

Copyright, 1908, by American School of Correspondence

all points on the surface of the earth, it follows that the foot-pound, or *gravitation-measure* of work, is not a constant unit. Its variation, however, is so small as to be negligible for ordinary purposes; and, being much simpler than the theoretically accurate units which must occasionally be employed in scientific investigation, it has remained in very general use. Thus the work done in raising 20 pounds of water through a height of 1 foot, or 1 pound of water through a height of 20 feet, or 5 pounds of water through a height of 4 feet, is said to be 20 foot-pounds.

3. **Power.** In the preceding definition, the element of *time* was not considered; thus, in the above example, 20 foot-pounds of work were done, whether the indicated operation took one minute to perform or extended over a period of one hour, or longer. The term *power* is defined as the amount of energy that can be exerted, or work done, *in a given time*.

4. **Unit of Power.** For industrial purposes, the unit most commonly employed is the *horse-power* (h.p.), which represents the capacity to perform 33,000 foot-pounds of work in one minute, or 550 foot-pounds of work in one second; it thus indicates the *rate of work*.

Example 1. A pump raising 7,500,000 gallons of water* in 10 hours to an elevated tank 50 feet high, is performing:

$$\frac{7,500,000 \times 62.5 \times 50}{7.5} = 3,125,000,000 \text{ ft.-lbs. of useful work; or,}$$

$$\frac{3,125,000,000}{10 \times 60} = 5,208,333 \text{ ft.-lbs. per minute,}$$

which is equivalent to:

$$\frac{5,208,333}{33,000} = 157.8 \text{ h.p.}$$

This amount of horse-power is the rate of work which, in the example above, must be continued for 10 hours in order to raise the total quantity of water. The entire problem may be conveniently performed in one operation, thus:

$$\frac{7,500,000 \times 62.5 \times 50}{7.5 \times 10 \times 60 \times 33,000} = 157.8 \text{ h.p.}$$

5. **Energy.** The amount of energy existing in any agent is measured by the quantity of work it is able to do; *energy* and *work*

---

*One cubic foot of water weighs 62.5 lbs. and contains 7.5 gallons (approximately).

are therefore measured by the same unit. "When energy is exerted, work is done against resistance." As usually stated in Theoretical Mechanics, energy may exist as *potential energy*—energy of position; or *kinetic energy*—energy of motion; or partly in one form, and partly in the other. Thus (see Fig. 1) a cannon-ball weighing W pounds, located in an elevated position h feet above any plane of reference, possesses Wh foot-pounds of potential energy with respect to that plane, by virtue of its position. If it be allowed to fall to the plane, it will, at its lowest point, theoretically have acquired a velocity of $v(=2gh)$ feet per second, and will therefore, at that level, possess kinetic energy to the amount of $W\dfrac{v^2}{2g}$ ($=Wh$) foot-pounds by reason of its motion. Further, if we analyze the conditions at some intermediate plane h_1 feet below its original position, and h_2 feet above the lower level, we shall find that the ball has acquired at this point a velocity of v_1 ($=\sqrt{2gh_1}$) feet per second, and therefore possesses kinetic energy to the amount of $W\dfrac{v_1^2}{2g}$ ($=Wh_1$) foot-pounds due to its

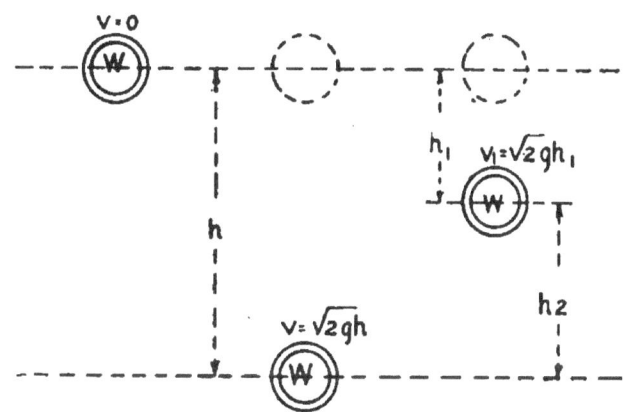

Fig. 1. Illustrating Relation between Potential and Kinetic Energy.

motion; but, by reason of its position h_2 feet above the lower plane, it still possesses Wh_2 foot-pounds of potential energy; consequently, with respect to the lower plane, the ball possesses a total energy represented by $W\left(\dfrac{v_1^2}{2g} + h_2\right) = W(h_1 + h_2) = Wh = W\dfrac{v^2}{2g}$ foot-pounds. Thus potential and kinetic energies are mutually convertible, theoretically without loss; practically, more or less energy will be transformed into heat during the conversion, and dissipated. But the great principle of the Conservation of Energy teaches that the *total quantity* of energy existing, or stored in the ball in any position, is theoretically a constant quantity.

6. Pressure-, Velocity-, and Gravity-Head. In hydraulic work, because of the nature of the medium dealt with—water being considered in this connection a perfect fluid, and incompressible—and because of the character of the problems presented, it is customary and convenient to consider the energy of water as capable of existing in three forms—*Pressure*, *Velocity*, and *Gravity*. Thus, in Fig. 2, with the conditions as represented (see also "Hydraulics," page 34), if the valve at D be closed, the water will rise in tube CC (called a *piezometer tube*) to the same level EF as that existing in the reservoir, and the pressure in the pipe at C will be represented by the head h

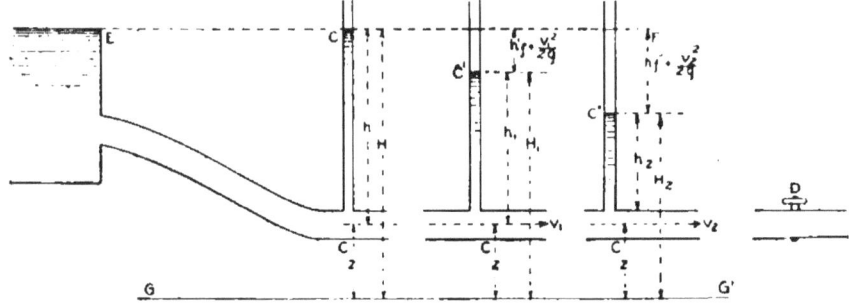

Fig. 2. Illustrating Relations of Pressure-, Velocity-, and Gravity-Head.

feet. Now, if the valve at D be partially opened, so that there is some velocity of flow v_1, in the pipe at section C, the column of water in the tube CC will sink to some lower level, as CC^n, and the pressure in the pipe at C will be that due to the head h_1 feet. Similarly, if the valve be now completely opened, so that the velocity of flow v_2, in the same section, becomes greater than v_1, the column of water in the tube will sink still lower, as CC''', indicating a pressure in the pipe at C represented by the head h_2 feet. If the loss of head in friction, etc., in the two cases of flow indicated above be respectively represented by h'_f and h''_f, the important relations existing are clearly shown in this diagram. It is evident that at the end of the pipe, where the water discharges freely into the air, no pressure-head exists, all the energy possessed by the issuing water being kinetic.

7. Total Head. Now let GG^1 represent any horizontal plane of reference—for example, the level of the tail-race water in a hydraulic power plant. With reference to this plane, the total effective head existing in the pipe at the section C, is:

(a) *For the case of no flow—*
$$z + h = H \text{ feet};$$
(b) *For the case of partial flow—*
$$z + h_1 + \frac{v_1^2}{2g} = H - h'_f \text{ feet};$$
(c) *For the case of full flow—*
$$z + h_2 + \frac{v_2^2}{2g} = H - h''_f \text{ feet}.$$

The distance z may be called the *gravity-head* (it corresponds to the head in potential energy already referred to); $\frac{v_1^2}{2g}$ and $\frac{v_2^2}{2g}$ are properly termed the *velocity-heads* (they correspond to the heads in kinetic energy already explained); h_1 and h_2 are known as the *pressure-heads* (see "Hydraulics," Article 6); h'_f and h''_f represent the heads lost in overcoming the various resistances to flow, principally friction in the pipe for the usual cases; but in the general case they include losses of head due to entrance, valves, curves, etc. (see "Hydraulics," Articles 28 and 34).

8. Energy per Pound of Water. The quantities stated above as number of feet in (a), (b), and (c) may be understood in another sense. Each may represent the total number of foot-pounds of energy existing *in every pound of water* in, or passing through, the pipe at section C; thus,

(a) $z + h = H$ foot-pounds per pound of water

(b) $z + h_1 + \frac{v_1^2}{2g} = H - h'_f$ " " " " " "

(c) $z + h_2 + \frac{v_2^2}{2g} = H - h''_f$ " " " " " "

9. Total Energy. Now suppose W_1 and W_2 pounds of water per second respectively to pass the section C in the two cases of flow considered; then, with respect to the plane GG^1, the total energy of the water as it passes this section is, for the one case:

(b) $W_1 (z + h_1 + \frac{v_1^2}{2g})$ foot-pounds;

and for the other:

(c) $W_2 (z + h_2 + \frac{v_2^2}{2g})$ foot-pounds;

and these expressions represent, for the two cases considered, the *total amount of energy* possessed by the water, *with respect to the plane GG^1*, and theoretically capable of being delivered to a machine or motor, by the descent of the water from the upper level EF to the lower level GG^1.

Where the water issues freely into the air from the extremity of the pipe, or through a nozzle at the end, no pressure exists; therefore, in the expression corresponding to (b) or (c), above, for such section, the term representing pressure-head disappears, leaving the two terms indicating gravity-head and velocity-head.

Further, if the plane of reference passes through the center of the end of the pipe or nozzle opening, the term representing gravity-head also disappears, leaving the velocity-head alone to indicate the energy of the stream at this point.

It is usually more convenient to express the sum of gravity-head and pressure-head in a single term: thus, $z + h_1 = H_1$; and $z + h_2 = H_2$; here H_1 and H_2 may be called the *piezometer heights*.

10. Efficiency. The efficiency of any apparatus for utilizing the kinetic energy of moving water, or the potential energy of still water, is *the ratio of the amount of work given out by the apparatus to the amount of work delivered to it*; or, as it is sometimes stated, it is *the ratio of the useful work to the theoretic energy*. This topic will be treated more fully in a later article; for the present, if e represent the efficiency of a motor, then,

$$e = \frac{\text{Foot-pounds or horse-power } given\ out\ by \text{ motor}}{\text{Foot-pounds or horse-power } delivered\ to \text{ motor}}.$$

As will be seen later, the denominator does not represent the full theoretic energy of the waterfall, since more or less of this energy must be utilized in overcoming the resistances encountered in conducting the water to the motor.

Example 2. A motor is operated by a stream of water discharged through a 2-foot pipe with a velocity of 10 feet per second. The motor gives out at its shaft 4.4 horse-power. What is the efficiency of the motor?

$$\frac{3.1416 \times 10 \times 62.5 \times 100}{550 \times 64.4} = 5.5 \text{ horse-power delivered to motor}$$

$$e = \frac{4.4}{5.5} = 80 \text{ per cent efficiency.}$$

Example 3. A small turbine wheel using 100 cubic feet of water per minute under a head of 45 feet, is found to give 6 horse-power. What is the efficiency of the wheel?

6 Horse-Power = $6 \times 33,000 = 198,000$ ft.-lbs. per min.

$$e = \frac{198,000}{100 \times 62.5 \times 45} = 70.4 \text{ per cent efficiency.}$$

WATER-POWER DEVELOPMENT 7

Theoretic Efficiency. If the efficiency of the motor actuated by the water were 100 per cent, it would give out at its shaft, as useful work, the same number of foot-pounds that were delivered to it. It is also interesting to note that if the efficiency of the hydraulic parts of the plant were 100 per cent—that is, if there were no hydraulic losses of head—the total energy of the water (see Fig. 2) represented by the total head H feet, or H foot-pounds per pound of water, would be available; and, if operating a motor of 100 per cent efficiency, the total energy of the water would be given out as useful work at the shaft of the motor. In practice these ideal conditions can never be fully realized, for there are certain hydraulic and mechanical losses of energy, which, while they may be reduced to the lowest limits by means of proper design, nevertheless, cannot be entirely eliminated.

Example 4. A pond containing 2,000,000 cubic feet of water is at an average elevation of 50 feet above the lower level. How much potential energy does this theoretically represent at the lower level?

$$2,000,000 \times 62.5 \times 50 = 6,250,000,000 \text{ ft.-lbs.}$$

If this water is fed to a small motor at the rate of 100 cubic feet per minute, what horse-power does this represent, and how long may the motor be operated?

$$\frac{100 \times 62.5 \times 50}{33,000} = 9.5 \text{ h.p.}$$

$$\frac{2,000,000}{100 \times 60 \times 24} = 13\frac{7}{8} \text{ days, or 13 days 21 hours.}$$

Assuming that the motor has an efficiency of 75 per cent, how much power may be taken off at its shaft?

$$9.5 \times .75 = 7.1 \text{ h.p.}$$

Example 5. The discharge of a stream is 1,000 cubic feet per second; its mean velocity is 3 feet per second. What horse-power does this represent?

$$\frac{1,000 \times 62.5 \times (3)^2}{550 \times 64.4} = 1,588.1 \text{ h. p.}$$

Example 6. Water issues from a nozzle at the rate of 50 feet per second; the area of the nozzle opening is 0.1 square foot. How many foot-pounds of kinetic energy does this represent? How many horse-power? If this jet operates a motor of 80 per cent efficiency, what horse-power will the motor actually yield?

$$0.1 \times 50 \times 62.5 \times \frac{(50)^2}{64.4} = 21,125 \text{ ft.-lbs. per second.}$$

$$\frac{21,125}{550} = 22 \text{ h.p.}$$

$$22 \times .80 = 17.6 \text{ h.p.}$$

11. Pipe End with Nozzle. *Pressure at Base of Nozzle.*

For many purposes—as in hydraulic mining, in the operation of certain types of water motor (described later), and at the extremity of fire-hose—water is delivered at considerable velocity through a nozzle attached to the end of a pipe. It is therefore desirable to develop a formula for velocity of flow, and quantity of discharge, for such cases.

If the pressure-head h_1 (Fig. 3) at the entrance or base of a *smooth* nozzle be observed, either by a piezometer tube or by a pressure

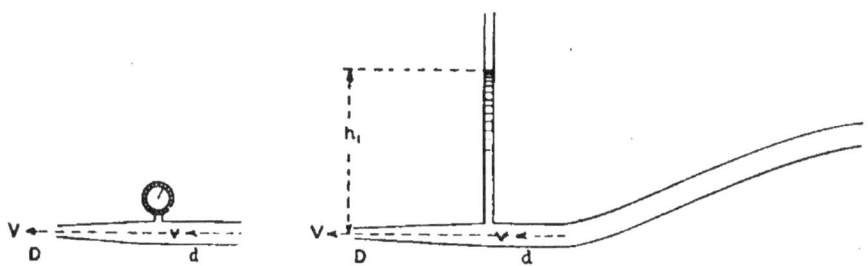

Fig. 3. Pipe with Nozzle Attachment.

gauge, then, since the nozzle velocity V is a consequence of the pressure-head h_1 and the velocity-head $\frac{v^2}{2g}$ of the water in the pipe approaching the nozzle with a velocity of v feet per second, the real or effective head on the nozzle is $h_1 + \frac{v^2}{2g}$; the theoretic velocity from the nozzle is:

$$V = \sqrt{2g\left(h_1 + \frac{v^2}{2g}\right)};$$

and the actual velocity is:

$$V = c_1 \sqrt{2g\left(h_1 + \frac{v^2}{2g}\right)},$$

in which c_1 denotes the coefficient of velocity, which, for smooth nozzles, is the same as the coefficient of discharge. In these equations, h_1 is expressed in feet; V and v in feet per second. Let D and d be the diameters, in feet, of the nozzle and pipe respectively.

Since the Discharge q = Area × Velocity,

$$q = \frac{\pi D^2}{4} V = \frac{\pi d^2}{4} v;$$

therefore,

WATER-POWER DEVELOPMENT

$$v = \left(\frac{D}{d}\right)^2 V.$$

Substituting this value of v in the equation above, and solving for V, there results:

$$V = \sqrt{\frac{2gh_1}{\left(\frac{1}{c_1}\right)^2 - \left(\frac{D}{d}\right)^4}} \quad \ldots \ldots \ldots (1)$$

in feet per second; and the discharge (area $\times V$) is:

$$q = 0.7854\, D^2 \sqrt{\frac{2gh_1}{\left(\frac{1}{c_1}\right)^2 - \left(\frac{D}{d}\right)^4}} \quad \ldots \ldots (2)$$

in cubic feet per second; and the velocity-head of the issuing jet is:

$$\frac{V^2}{2g} = \frac{h_1}{\left(\frac{1}{c_1}\right)^2 - \left(\frac{D}{d}\right)^4} \quad \ldots \ldots \ldots (3)$$

In many cases it is common to read the pressure at the base of the nozzle in pounds per square inch; then h_1 (in feet) equals $2.304\, p_1$ (in pounds per square inch); and the discharge is frequently stated in gallons per minute; making these substitutions in Equation 2, above, we have:

$$q = 29.83\, D^2 \sqrt{\frac{p_1}{\left(\frac{1}{c_1}\right)^2 - \left(\frac{D}{d}\right)^4}} \quad \ldots \ldots (4)$$

in gallons per minute.

Example 7. The pressure-gauge at the base of a smooth 1¼-inch nozzle reads 80 pounds per square inch; compute the velocity and discharge from the nozzle, the velocity-head of the issuing stream, and the mean velocity in the pipe, if the latter be 2½ inches in diameter. Assume 0.97 as the value of the coefficient.

Substituting the given numerical values in Equation 1, we have:

$$V = \sqrt{\frac{64.4 \times (2.304 \times 80)}{\left(\frac{1}{.97}\right)^2 - \left(\frac{1}{2}\right)^4}} = 38.5 \text{ foot per second.}$$

$$q = \text{Area} \times V = \frac{0.7854 \times 1.25}{144} \times 38.5 = 0.33 \text{ cubic foot per second.}$$

$$\frac{V^2}{2g} = \frac{(38.5)^2}{64.4} = 22.7 \text{ feet.}$$

$$v = \left(\frac{D}{d}\right)^2 V = \tfrac{1}{4} \times 38.5 = 9.6 \text{ feet per second.}$$

What horse-power does this represent?

$$\frac{0.33 \times 62.5}{550} \times 22.7 = 0.85 \text{ h.p}$$

With a motor of 80 per cent efficiency, how much useful work will be obtained?

$$0.85 \times 0.80 = 0.68 \text{ h.p.}$$

12. Pipe Line with Nozzle. In Fig. 4, let h be the total head on the end of the nozzle, D its smaller diameter in feet, and V the velocity of the issuing stream in feet per second. Let d and v be the corre-

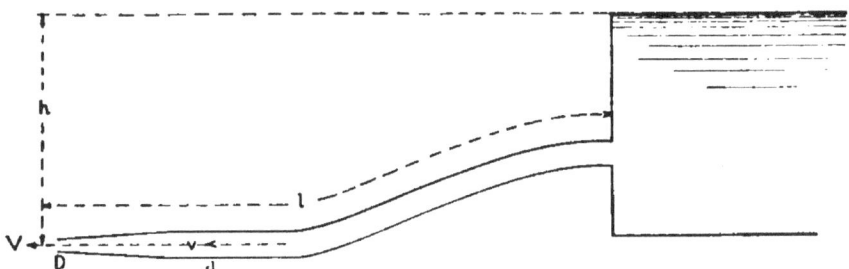

Fig. 4. Loss of Head in Pipe and Nozzle.

sponding quantities for the pipe or hose, and l its length in feet. Of the total available head h on the end of the nozzle, only $\dfrac{V^2}{2g}$ remains; so that $h - \dfrac{V^2}{2g}$ represents the head lost or dissipated in overcoming various resistances to flow, from the reservoir to the tip of the nozzle. This lost head consists of several parts (see "Hydraulics," Article 34); we may therefore write:

$$h - \frac{V^2}{2g} = \left\{ \left(\frac{1}{c}\right)^2 - 1 \right\} \frac{v^2}{2g} + \frac{8gl}{C^2 d} \frac{v^2}{2g} + m\frac{v^2}{2g} + n\frac{v^2}{2g} + m'\frac{V^2}{2g} \quad \ldots (7)$$

in which,

$\left\{ \left(\dfrac{1}{c}\right)^2 - 1 \right\} \dfrac{v^2}{2g}$ = Loss of head at entrance; $\dfrac{8gl}{C^2 d} \dfrac{v^2}{2g}$ = Head lost in friction in the pipe (see "Hydraulics," Articles 28 and 36); $m\dfrac{v^2}{2g}$ = Head lost in bends and curves; $n\dfrac{v^2}{2g}$ = Head lost by the passage of the water through valves and gates; and, lastly, $m'\dfrac{V^2}{2g}$ = Head lost in passing through the nozzle.

The equation for the value of m' assumes a form similar to that for entrance loss into a pipe:

$$m' = \left\{ \left(\frac{1}{c_1}\right)^2 - 1 \right\},$$

in which c_1 is the coefficient of velocity, which, for smooth nozzles, is the same as the coefficient of discharge; its value may be taken as 0.97 for such nozzles, with the small diameter between $\tfrac{3}{4}$ inch and $1\tfrac{1}{2}$ inches, under ordinary range of pressures.

Since, in steady flow, the velocities v and V are inversely proportional to the areas of the corresponding cross-sections,

$$V = v\left(\frac{d}{D}\right)^2$$

Inserting this value of V in Equation 5, and solving for v, there results:

$$v = \sqrt{\frac{2gh}{\left\{\left(\frac{1}{c}\right)^2 - 1\right\} + \frac{8gl}{C^2 d} + m + n + \left(\frac{1}{c_1}\right)^2 \left(\frac{d}{D}\right)^4}} \quad\quad(6)$$

for the velocity of flow in the pipe, in feet per second.

The velocity and discharge from the nozzle are then:

$$V = \left(\frac{d}{D}\right)^2 v, \quad\quad\quad\quad\quad\quad(7)$$

and,

$$q = \frac{1}{4}\pi D^2 V \quad\quad\quad\quad\quad\quad(8)$$

In many cases the sum of the losses at entrance, through valves and gates, and around bends and curves, is sufficiently small, in comparison with the loss in pipe friction, to be negligible; in such cases, Equation 6 reduces to

$$v = \sqrt{\frac{2gh}{\frac{8gl}{C^2 d} + \left(\frac{1}{c_1}\right)^2 \left(\frac{d}{D}\right)^4}} \quad\quad(9)$$

Example 8. A smooth nozzle with a small diameter of 1 inch is attached to a 3-inch pipe 1,500 feet long; the tip of the nozzle is 64 feet below the surface of the water in an elevated reservoir. Assume $C = 100$, and determine the velocity of flow in the pipe, and through the nozzle. Find also the discharge, and the efficiency of the pipe and nozzle.

Since in this case the entrance loss is relatively small, because the pipe is long in comparison with its diameter, and therefore pipe friction is relatively large, Equation 9 may be used:

$$v = \sqrt{\frac{64.4 \times 64}{\frac{8 \times 32.2 \times 1,500}{(100)^2 \times 0.25} + \left(\frac{1}{.97}\right)^2 \left(\frac{3}{1}\right)^4}} = 4.14 \text{ feet per second,}$$

for the velocity of flow in the pipe.

$$V = v\left(\frac{d}{D}\right)^2 = 4.14 \times 9 = 37.26 \text{ feet per second,}$$

for the velocity of the jet issuing from the nozzle.

$$q = \frac{\pi d^2}{4} v = \frac{3.1416 \times \left(\frac{1}{4}\right)^2}{4} \times 4.14 = 0.20 \text{ cu. ft. per second.}$$

The energy of the jet is:

$$W \frac{V^2}{2g} = \frac{.20 \times 62.5 \times (37.26)^2}{64.4} = 269.5 \text{ ft.-lbs. per second.}$$

The theoretic energy is:

$$Wh = .20 \times 62.5 \times 64 = 800 \text{ ft.-lbs. per second.}$$

The efficiency of pipe and nozzle, therefore, is:

$$\frac{269.5}{800} = 33.7 \text{ per cent.}$$

13. If, under the conditions just stated, we suppose the nozzle removed, the last term in the denominator of Equation 9 will disappear, and the equation will assume the form:

$$v = \frac{C}{2}\sqrt{\frac{hd}{l}} = C\sqrt{r\frac{h}{l}} = C\sqrt{rs} \quad \ldots \ldots (10)$$

which is Equation 30 in "Hydraulics," for the case of a pipe of uniform diameter; or Equation 33, for flow in open channels.

14. Equation 7, taken in connection with Equation 6 or its simpler form, Equation 9, shows that the smaller the nozzle diameter compared with that of the pipe, within ordinary practical limits, the greater will be the nozzle *velocity*; but the greatest *discharge* will occur (Equation 8) when the nozzle diameter is as large as possible; that is, when it is equal to the pipe diameter—in other words, when there is no nozzle attached.

15. **Relation of Pipe and Nozzle Diameters.** When the object of attaching a nozzle to a pipe is to utilize the velocity-head of the issuing jet $\left(= \frac{V^2}{2g}\right)$ without regard to the quantity of water discharged, a large pipe and a relatively small nozzle should be employed. When the object is to obtain as large a discharge as possible, no nozzle should be used, and the pipe should be as large as practical considerations will warrant. When the object is to utilize the energy of the jet in producing power by means of a water-motor, in which case both velocity-head and quantity of discharge are concerned, there is a definite relation existing between the diameters of nozzle and pipe that will render this a maximum.

16. Maximum Power Derivable from Nozzle Jet. From Equations 9 and 7, we derive:

$$V = \sqrt{\frac{2gh}{\frac{Sfl}{C^2 d}\left(\frac{D}{d}\right)^4 + \left(\frac{1}{c_1}\right)^2}} \quad \ldots \ldots (11)$$

Then, if w be the weight in pounds of a cubic foot of water, we have, for the theoretical kinetic energy of the issuing jet in foot-pounds per second (weight of discharge in pounds per second × velocity-head):

$$K = w\frac{1}{4}\pi D^2 V \frac{V^2}{2g} = \frac{w\pi D^2 V^3}{8g} \quad \ldots \ldots (12)$$

Substituting in this equation the value of V above (Equation 11), and ascertaining, by the procedure usually adopted in such cases (differential calculus), the value of D to render K a maximum, we obtain:

$$D = 4\left(\frac{C^2 d^5}{g c_1^2 l}\right)^{\frac{1}{4}}, \quad \ldots \ldots (13)$$

which is a formula for diameter of nozzle in terms of diameter and length of pipe (all in feet) to produce the maximum kinetic energy of the jet issuing from the nozzle.

With a nozzle of this diameter, the velocity of the issuing jet is obtained by placing the value of D from Equation 13 in Equation 11, with the result:

$$V = 2c_1\sqrt{\frac{gh}{3}} = c_1\sqrt{2g\left(\frac{2}{3}h\right)} = 0.816 c_1 \sqrt{2gh} \quad (14)$$

Since the value of c_1 for ordinary cases is about 0.97, it may be said that the nozzle velocity necessary to produce the *maximum power* is about 80 per cent of the theoretic velocity due to the actual static head on the nozzle tip.

17. Relation between Total Head and Friction Head for Maximum Power. The relation expressed by Equation 14 leads to some interesting conclusions. Since $V = .80 \sqrt{2gh}$ for maximum power, $\frac{V^2}{2g} = .64 h$; therefore, since the total head is h, $.36h$ must be used in overcoming pipe and nozzle resistance, to give the most advantageous velocity for power purposes. Again, omitting nozzle resistance (as represented by c_1), $\frac{V^2}{2g} = .667h$; therefore $.333h$ must be used in overcoming pipe friction alone. That is to say, with the conditions arranged to furnish maximum power, $\frac{1}{3}$ of the total static head on the nozzle tip is being used to overcome pipe friction, and the remaining $\frac{2}{3}h$ is transformed into the velocity-head of the issuing stream after due deduction or allowance for nozzle resistance. The second value of V (Equation 14) shows this directly. If no nozzle is attached, therefore, the conditions for maximum power obtain when $\frac{1}{3}$ the total static head is used in overcoming pipe friction, the remaining $\frac{2}{3}$ of the head being available as velocity-head, or as pressure-head, or partly in one form and partly in the other.

18. Usually the discharge in cubic feet per second (q) is known; then, by simple substitution (Equations 8 and 14), the values for maximum work are:

$$D = \left(\frac{12 q^2}{\pi^2 c_1^2 gh}\right)^{\frac{1}{4}} \quad \ldots \ldots \ldots \ldots (15)$$

and, from Equations 13 and 15:

$$d = 2\left(\frac{6 q^2 l}{\pi^2 C^2 h}\right)^{\frac{1}{5}} \quad \ldots \ldots \ldots \ldots (16)$$

in which D and d are the diameters in feet of nozzle tip and pipe to furnish maximum power. Being stated in terms of q, l, and h, these equations are occasionally the most convenient to use in solving problems.

Example 9. By damming a stream, an impounding reservoir was created, capable of supplying uniformly 5.92 cubic feet of water per second to a powerhouse below. The nozzle tip is to be 590 feet below the average water level in the reservoir; the length of pipe is 6,000 feet from reservoir to nozzle; the pipe being of riveted steel, and making due allowance for deterioration of surface with age, C was assumed to have the low value 83. What size pipe and nozzle should be used to give the maximum power? What will be the nozzle velocity? What horse-power will be developed at the nozzle? What efficiency does this represent for pipe and nozzle? What power may be derived from a wheel of 75 per cent efficiency, driven by the jet? What is the efficiency of the whole system?

From Equation 16:

$$d = 2 \left\{ \frac{6 \times (5.92)^2 \times 6,000}{(3.1416)^2 \times (83)^2 \times 590} \right\}^{\frac{1}{5}} = 1 \text{ foot, pipe diameter.}$$

From Equation 15:

$$D = \left\{ \frac{12 \times (5.92)^2}{(3.1416)^2 \times (.97)^2 \times 32.2 \times 590} \right\}^{\frac{1}{4}} = 2.67 \text{ inches, nozzle diameter,}$$

or Equation 13 may be used to determine D.

From Equation 14:

$$V = 0.816 \times 0.97 \sqrt{64.4 \times 590} = 152 \text{ feet per second, nozzle velocity.}$$

$$\text{Horse-power} = \frac{WV^2}{2g \times 550} = \frac{5.92 \times 62.5 \times (152)^2}{64.4 \times 550} = 241 \text{ h.p.}$$

$$\text{Theoretic horse-power} = \frac{Wh}{550} = \frac{5.92 \times 62.5 \times 590}{550} = 397 \text{ h.p.}$$

$$\text{Efficiency} = \frac{241}{397} = 61 \text{ per cent (nearly).}$$

Useful work from wheel = $241 \times .75 = 181$ h.p.

Efficiency of whole system = $.75 \times .61 = 46$ per cent (or $\frac{46}{100}$).

19. **Multiple Nozzles.** Sometimes an impulse wheel is driven by means of jets issuing from two or more nozzles of the same or of different diameters. Then, for maximum power, the sum of the areas of the several nozzles must equal the area corresponding to D, as computed for a single nozzle, on the assumption that the nozzle tips are at substantially the same level, and that the coefficient c_1 has the same value for each. Thus, if there be two nozzles with diameters D_1 and D_2,

$$D_1^2 + D_2^2 = \frac{1}{4}\left(\frac{C^2 d^5}{g c_1^2 l}\right)^{\frac{1}{2}} = \frac{C d^2}{4 c_1}\sqrt{\frac{d}{gl}} \quad \ldots (17)$$

One diameter, as D_1, may be assumed, and the other computed from the above relation.

If the two nozzles are of equal diameter D_1,

$$D_1^2 = \frac{1}{8}\left(\frac{C^2 d^5}{g c_1^2 l}\right)^{\frac{1}{2}};$$

therefore,

$$D_1 = \frac{1}{2}\left(\frac{C^2 d^5}{4 g c_1^2 l}\right)^{\frac{1}{4}} \quad \ldots \ldots (18)$$

If the value of D for one nozzle has already been determined, then, for two nozzles of equal diameter D_1, from the relation stated above,

$$\frac{2 \pi D_1^2}{4} = \frac{\pi D^2}{4};$$

therefore,

$$D_1 = \frac{D}{\sqrt{2}} \quad \ldots \ldots (18\text{a})$$

With three or more nozzles, of the same or of different diameters, the relation of areas stated above will furnish a means of readily determining the diameters. Thus, for three nozzles of equal diameter D_1,

$$D_1 = \frac{D}{\sqrt{3}} \quad \ldots \ldots (18\text{b})$$

If the discharge q is known, an analysis similar in all respects to that above will give, in place of Equation 17:

$$D_1^2 + D_2^2 = \frac{2q}{\pi c_1}\sqrt{\frac{3}{gh}}; \quad \ldots\ldots\ldots (19)$$

and, in place of Equation 18:

$$D_1 = \left(\frac{3q^2}{\pi^2 c_1^2 gh}\right)^{\frac{1}{4}}, \quad \ldots\ldots\ldots\ldots (20)$$

which will prove more convenient for use in some problems.

Example 10. If, in example 9, two nozzles of equal diameter were required, the diameter of each nozzle could be determined directly from Equation 18; or more simply, from Equation 18a, since the value of D has already been found:

$$D_1 = \frac{D}{\sqrt{2}} = \frac{2.67}{1.41} = 1.9 \text{ inches for each nozzle.}$$

If three equal nozzles were required, then, from Equation 18b:

$$D_1 = \frac{D}{\sqrt{3}} = \frac{2.67}{1.73} = 1.5 \text{ inches for each nozzle.}$$

IMPULSE, REACTION, AND DYNAMIC PRESSURE

20. Impulse and Reaction of Water in Motion. Let W be the number of pounds of water discharged per second from an orifice, pipe, or nozzle, or flowing in a stream, with a uniform velocity of v feet per second; then,

$$F = W\frac{v}{g} \text{ pounds} \ldots\ldots\ldots\ldots (21)$$

is called the *impulse* of the moving water. It may be regarded as a continuous pressure in the direction of motion; and it will be exerted as such upon a surface placed in the path of the jet or stream, with an intensity varying with the conditions, and ranging to the maximum value F, above. The *reaction*, or *back-pressure*, is equal in value to the impulse, but opposite in direction. For example, if a vessel containing water be freely suspended at A (Fig. 5), and water be allowed to flow out through an orifice at B, the pressure due to the head of water h causes W pounds of water per second to be discharged with the velocity v ($=$ theoretically $\sqrt{2gh}$) feet per second. In the direction of the jet, the impulse produces motion; in the opposite direction, it produces an equal back-pressure (action and reaction being equal in amount and opposite in direction), causing the vessel to swing to the right. The first of these forces is the *impulse*, and the second is the *reaction* of the jet; and if a force R be applied as shown, of just sufficient intensity to prevent this motion of the vessel, its value is:

$$R = W\frac{v}{g} = F, \quad \ldots \ldots \ldots \ldots (22)$$

which is the reaction of the jet.

21. The impulse or reaction of a jet issuing from an orifice is double the hydrostatic pressure on the area of the orifice. For, if a is the area of the orifice, and w the weight of a cubic unit of water, the normal hydrostatic pressure on the area of the orifice when closed (see "Hydraulics," Article 6) is:

Hydrostatic pressure $= wah$ pounds.

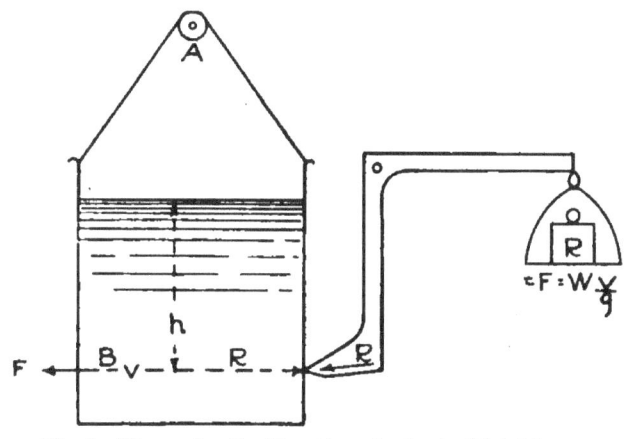

Fig. 5. Measuring the Reaction of a Jet by Weighing.

When the orifice is opened, the weight of the discharge per second (see "Hydraulics," Article 18) is theoretically $W = wav$; hence,

$$F = R = W\frac{v}{g} = wav\frac{v}{g} = \frac{2wav^2}{2g} = 2\,wah. \ldots (23)$$

This conclusion has been verified by many experiments (see Fig. 6).

Example 11. What must be the velocity of a jet of water 1 inch in diameter, issuing from a nozzle, in order that its impulse may be 100 pounds? What will be the discharge in cubic feet and in gallons per second?

$$F = \frac{Wv}{g} = \frac{wav^2}{g} = 100 \; ;$$

$$\therefore v = \sqrt{\frac{100 \times 32.2}{62.5 \times .0054}} = 97.7 \text{ foot per second.}$$

$$q = av = .0054 \times 97.7 = .53 \text{ cubic foot per second.}$$

$$.53 \times 7.5 = 4 \text{ gallons per second.}$$

22. **Dynamic Pressure of Water in Motion.** If a jet of water strike a stationary plane normally, it produces a dynamic pressure on that plane equal to the impulse of the jet; that is:

$$P = F = W\frac{v}{g}$$

If a jet moving with a velocity v_1 be retarded by a surface so that its velocity becomes v_2, without changing its direction, the impulse in the first case is:

$$F_1 = \frac{W v_1}{g};$$

and in the second case:

$$F_2 = \frac{W v_2}{g};$$

and the difference,

$$P = F_1 - F_2 = W\left(\frac{v_1 - v_2}{g}\right) \quad \ldots \ldots \ldots (24)$$

is a measure of the dynamic pressure which has been developed in

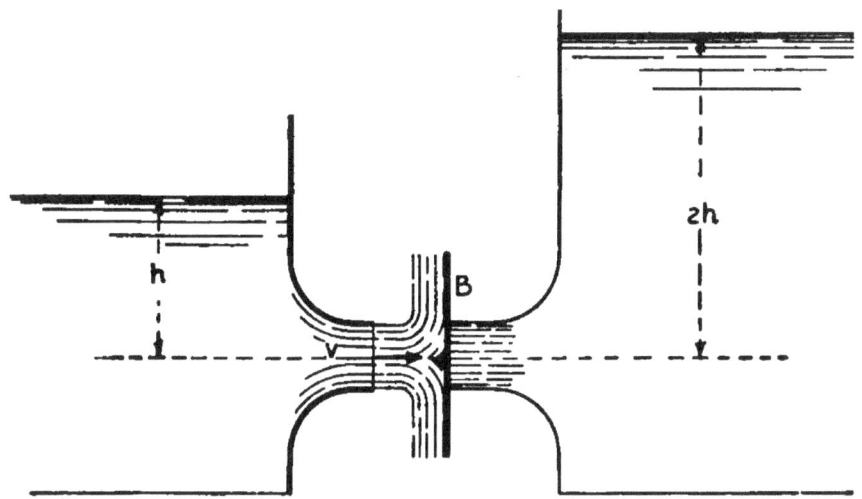

Fig. 6. Illustrating Relation between Impulse and Hydrostatic Pressure.

the direction of motion by the retardation of the velocity. If a jet of water impinge upon a stationary surface which changes its direction of motion without changing its velocity, a dynamic pressure is developed, its amount depending upon the velocity and the change in direction. In all cases this pressure is exerted upon the surface causing the retardation of velocity or change in direction of flow.

23. **Static and Dynamic Pressures.** *Dynamic pressure* must be clearly distinguished from *static pressure*, the laws governing in the two cases being entirely different. A static pressure due to a given head will cause a jet of water to be discharged from an orifice with a velocity proportional to the head; if this jet impinge upon a surface, a dynamic pressure will be exerted upon it, which may be equal to, greater than, or less than the static pressure due to the head,

depending upon the circumstances. Again, at any point below the surface of water, static pressure is exerted with equal intensity in all directions; dynamic pressure is exerted with different intensities in different directions.

24. **Definitions.** From a comparison of Equations 21 and 24, we may now define the *impulse* of a jet or stream of water as the dynamic pressure which it is capable of producing in the direction of its motion when its velocity in that direction is entirely destroyed. This may be accomplished by carefully deflecting the jet 90 degrees to its original path by means of a smooth surface, so that, no energy being dissipated in overcoming frictional or other resistances, the velocity of the water is not changed, but its component in the original direction is zero; and the *reaction* of a jet or stream of water may be defined as the backward dynamic pressure, in the line of mo-

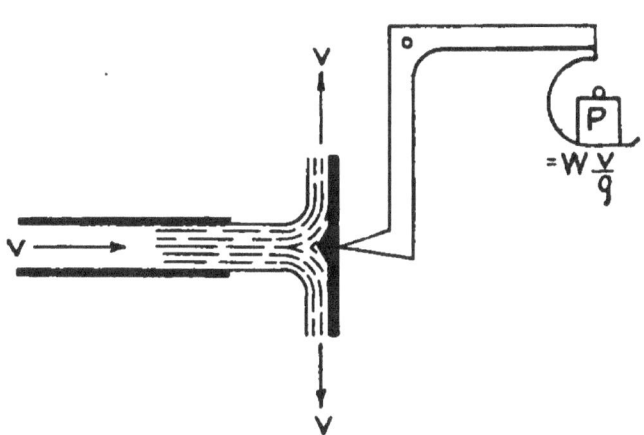

Fig. 7. Measuring Pressure of a Jet on a Plane Surface, by Weighing.

tion, which is exerted against a vessel out of which it issues, or against a surface away from which it moves.

25. **Laboratory Experiments on Impulse, Reaction, and Dynamic Pressure.** Fig. 5 shows how the reaction of a jet may be measured; the necessary weight in the scale pan to prevent motion of the vessel has been found to be very nearly:

$$R = F = \frac{Wv}{g} = 2wa\frac{v^2}{2g}.$$

Fig. 6 shows how the pressure due to the impulse of a jet may be made to balance the hydrostatic pressure due to twice the head causing the flow. B is a loose plate with surface carefully finished to fit the mouthpiece so as to prevent leakage. Fig. 7 illustrates a simple device for measuring by weighing the dynamic pressure exerted upon a surface by the impulse of a jet impinging upon and gliding over it, when its motion in the original direction has been entirely destroyed by being deflected 90 degrees. The result of the experiment is found to show very nearly that:

$$P = W\frac{v}{g} = 2wa\frac{v^2}{2g},$$

as theory requires.

Fig. 8 illustrates a case of dynamic pressure exerted upon a curved surface, due to both impulse and reaction, the former being due to the direct impact of the jet, the latter to the circumstance that the deflected stream leaves the surface in a direction which has a component of velocity parallel to the original path, but opposite in direction. Here experiment shows:

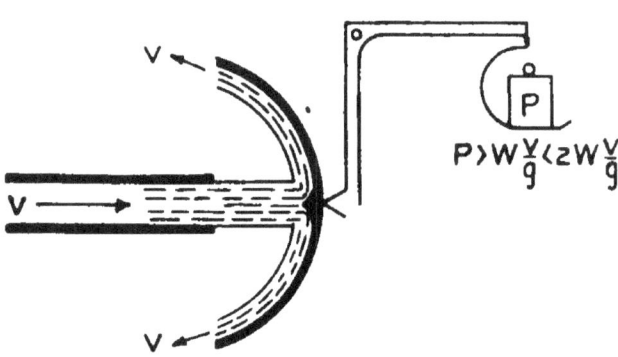

Fig. 8. Measuring Pressure from a Jet on a Curved Surface, by Weighing.

$$P > W\frac{v}{g} < 2W\frac{v}{g},$$

as theory requires.

Fig. 9 shows the case where the stream is deflected 180 degrees; that is, there is a complete reversal in the direction of motion; and we should expect the dynamic pressure exerted upon the surface to be equal to the sum of both impulse and reaction; namely,

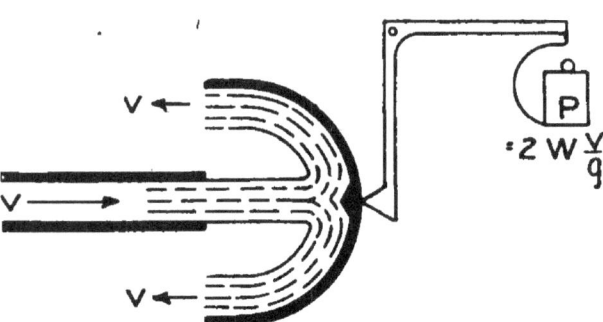

Fig. 9. Measuring Pressure from a Jet whose Direction is Completely Reversed.

$$P = F + R = 2F = 2W\frac{v}{g} = 4wa\frac{v^2}{2g},$$

which agrees quite closely with the results of laboratory experiments.

Example 12. In Fig. 7 the diameter of the tube is 1 inch; there is no contraction of the jet; and the discharge is .5 cubic foot per second. What is the velocity, and the dynamic pressure against the plane? What would be the dynamic pressure in the case represented by Fig. 9?

$$v = \frac{q}{a} = \frac{.5}{.0054} = 92.6 \text{ feet per second.}$$

$$P = W\frac{v}{g} = \frac{.5 \times 62.5 \times 92.6}{32.2} = 90 \text{ pounds.}$$

$$P = 2W\frac{v}{g} = 2 \times 90 = 180 \text{ pounds.}$$

FIXED SURFACES

26. Dynamic Pressures on Fixed Surfaces. When a stream of water impinges with a uniform velocity v on a smooth surface at rest, it glides over the surface and leaves it with the original velocity v, since there are supposed to be no frictional or other resistances, only its direction of motion being changed. The water, as it strikes the

Fig. 10. Illustrating Case of No Dynamic Pressure.

surface, exerts upon it an impulse F in the direction of the path of entry; as it leaves the surface, it exerts on it an equal reaction F, in a direction opposite to its path of exit (see Figs. 11 to 14). The dynamic pressure thus developed depends upon velocity v, and change of direction of stream (angle θ). The stream is assumed to be moving horizontally while in contact with the surface, so that its velocity is not affected by gravity.

27. Resultant Dynamic Pressure. From the principle of Composition of Forces (Mechanics), the resultant dynamic pressure upon a fixed surface struck by a jet may be readily found by constructing the parallelogram of the forces of impulse and reaction, as shown in Fig. 15, in which $ab = bc = F = R$; from which we deduce (Trigonometry) that the value of this resultant pressure is:

$$P_R = F\sqrt{2(1-\cos\theta)} = 2\sin\tfrac{1}{2}\theta \cdot W\frac{v}{g} \quad \textbf{(25)}$$

and that it makes an angle of $(90° - \tfrac{1}{2}\theta)$ with the original direction of the jet. Its line of action passes through the intersection of F and R, and it bisects the angle between them.

28. Dynamic Pressure Parallel to Initial Direction of Jet. This is simply the component of the Resultant Dynamic Pressure in the desired direction. From Fig. 16, this is found to be (Resolution of Forces) $ab = bc \cos(90 - \tfrac{1}{2}\theta)$; so that,

$$P_J = P_R \cos(90 - \tfrac{1}{2}\theta) = (1 - \cos\theta) W \frac{v}{g} \quad \ldots (26)$$

If, in this equation, $\theta = 0$, the stream glides over the surface without change of direction or retardation of velocity, and $P = 0$;

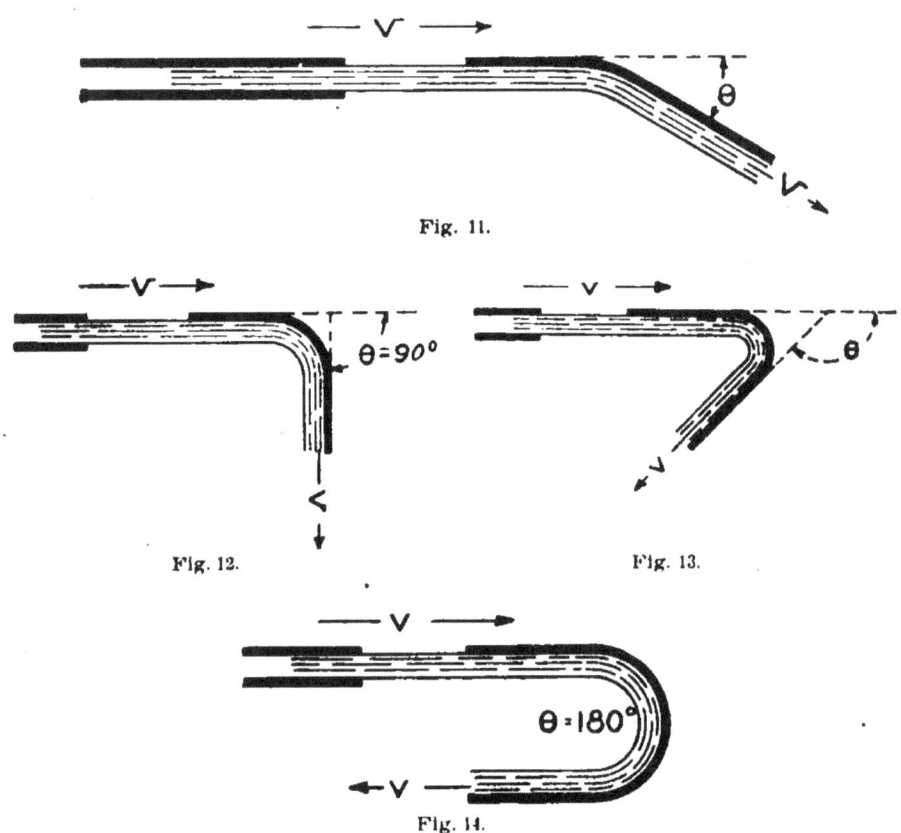

Illustrating Dynamic Pressure of Jet on Various Fixed Surfaces.

that is, no dynamic pressure is exerted upon the surface (see Fig. 10).

If $\theta = 90°$, $\cos\theta = 0$ (see Figs. 7 and 12), and therefore the dynamic pressure is:

$$P = F = W \frac{v}{g}$$

Here the escaping jet has no component of velocity normal to the surface; therefore the reaction has no influence on the pressure.

If $\theta = 180°$ (see Figs. 9 and 14), indicating a complete reversal in the direction of the stream, $\cos \theta = -1$; hence the dynamic pressure is:

$$P = 2F = 2W \frac{v}{g}.$$

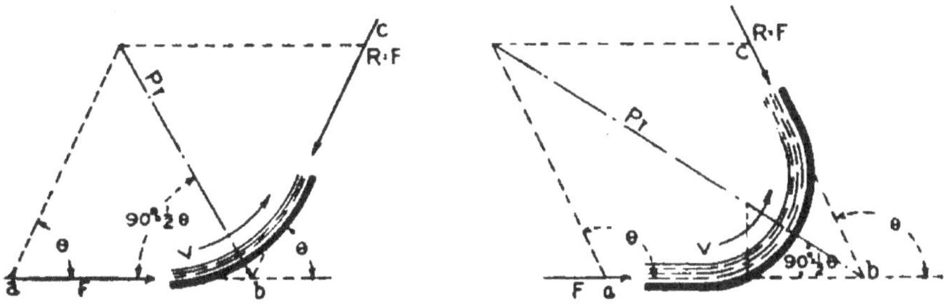

Fig. 15. Resultant Dynamic Pressure.

Here the pressure is a consequence of both impulse and reaction to their full amount.

29. Dynamic Pressure in Any Given Direction. It is frequently of importance to determine the dynamic pressure *in a given direction* exerted on a fixed surface by a stream of water. This may be ascertained by resolving the *resultant* dynamic pressure into its two components, parallel and at right angles to the required direction; the former represents the pressure in the required direction. Or the impulse and reaction may be separately resolved into their rectangular components, as above, and the algebraic sum taken of the two components parallel to the required direction. Thus, in Fig. 17, let it be required to find the dynamic pressure in a direction represented by the arrow x, which makes an angle a with the direction of the entering, and an angle θ with that of the departing stream. The components of the impulse and the reaction in the required direction, since $R = F$, are:

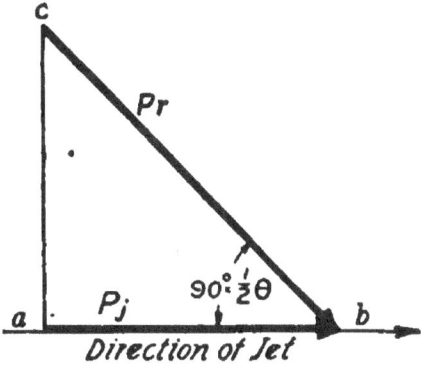

Fig. 16. Dynamic Pressure Parallel to Initial Direction of Jet.

$$P_1 = F \cos \alpha; \text{ and } P_2 = -F \cos \theta;$$

and therefore:

$$P = P_1 + P_2 = F(\cos \alpha - \cos \theta) = (\cos \alpha - \cos \theta) W \frac{v}{g} \quad (27)$$

If, in this general equation (27), $a = 0°$,

$$P = (1 - \cos \theta) W \frac{v}{g},$$

as in Equation 26.

If $a = 0°$, and $\theta = 90°$,

$$P = F = W \frac{v}{g}, \text{ as in Figs. 7 and 12}$$

If $a = 0°$, and $\theta = 180°$,

$$P = 2F = 2W \frac{v}{g}, \text{ as in Figs. 9 and 14.}$$

If $a = 0°$, and $\theta = 0°$, $P = 0$, as in Fig. 10.

Example 13. Let the jet of Problem 7 impinge tangentially upon the fixed curved vane of Fig. 15, with $\theta = 60°$. What is the resultant dynamic pressure upon the vane, in intensity and direction? What is the dynamic pressure in a direction parallel to the jet? What is the dynamic pressure in a direction making an angle of 30 degrees with the direction of the jet?

From Equation 25 and Problem 7:

$$P_R = 2 \sin \tfrac{1}{2} \theta \, W \frac{v}{g}$$

$$= 2 \times \tfrac{1}{2} \times \frac{.33 \times 62.5 \times 38.5}{32.2} = 24.7 \text{ pounds.}$$

From Equation 26:

$$P_J = (1 - \cos \theta) W \frac{v}{g}$$

$$= (1 - \tfrac{1}{2}) \frac{.33 \times 62.5 \times 38.5}{32.2} = 12.4 \text{ pounds.}$$

From Equation 27 (Fig. 17):

$$P = (\cos a - \cos \theta) W \frac{v}{g}$$

$$= (.866 - .500) \frac{.33 \times 62.5 \times 38.5}{32.2} = 4.5 \text{ pounds.}$$

30. Weight of Water Impinging. In all the preceding equations, W represents the weight of water in pounds per second impinging upon the surface; and, since the surface has in each case been assumed to be stationary, W is also the weight of water in pounds per second issuing from the nozzle or orifice, or flowing in the stream. It is to be clearly kept in mind that this statement is not necessarily true if the surface is supposed to move; as, for example, in the case of a jet impinging upon the vanes or blades of a water wheel. Such cases will be considered later.

31. Force and Work. It must also be clearly realized that the dynamic pressures are *forces*; they are not expressed in terms of *energy* or *work*; just as a weight resting upon a table produces *pressure* thereon, but does not perform *work*. A force must be exerted against a resistance through a definite distance, in order that work may be done; the weight may be allowed to move, and thereby compress a spring, for example, thus doing work. Similarly, the above pressures must be exerted against resistances over some definite distances, in order that work may be done. In general, if P is the

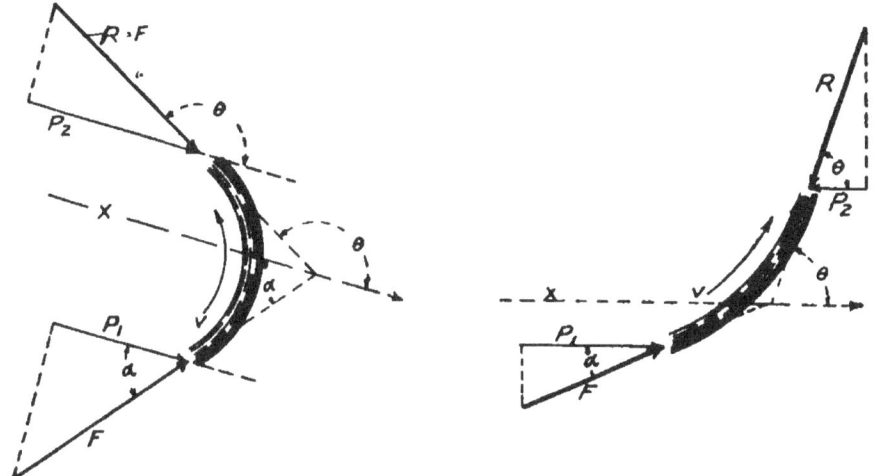

Fig. 17. Dynamic Pressure in Any Given Direction.

dynamic pressure on the surface in pounds, and if the surface is supposed to move a distance of u feet per second while overcoming some resistance, then,

$$\text{Work} = P \times u \text{ foot-pounds per second} \quad \ldots (28)$$

It is by reason of the dynamic pressures defined and explained above, produced by a retardation in velocity, or a change in direction of flow, that turbine wheels and other water-motors are able to transform the kinetic energy of moving water into useful work—such pressures being exerted over definite distances against resistances.

32. Losses of Energy. In the above discussion, no frictional or other losses of energy were considered. It is clear that if the surfaces are rough, or if the jet impinges on the surface in such a way as to produce "shock" or "eddies" or "foam," some of the original energy of the jet will be dissipated as heat, and the resulting pressures will be correspondingly reduced below the values indicated by the foregoing formulæ. These losses may be largely eliminated by having the surfaces smooth and properly curved, and by so directing the jet as to strike the surface tangentially.

ABSOLUTE AND RELATIVE VELOCITIES

33. Definitions. While all velocities are in reality relative, it is convenient to define *absolute velocity* as the rate of speed of a moving object with respect to the surface of the earth; and *relative velocity* as the rate of speed of a moving object with respect to another moving body—or as the velocity the object would appear to have to a person standing upon, and viewing it from, the second moving body. In the one case, velocity is measured from, or referred to, the earth,

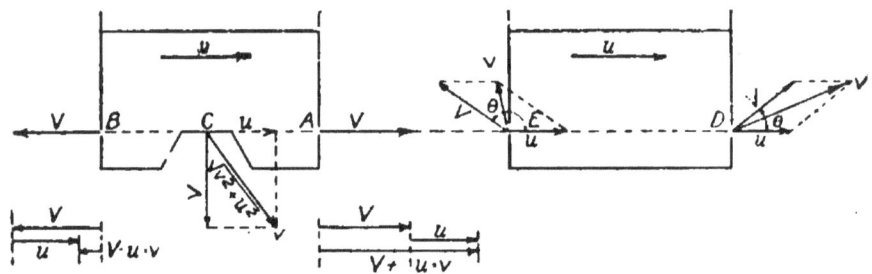

Fig. 18. Illustrating Absolute and Relative Velocities.

which is regarded as stationary; in the other case, the velocity is measured from, or referred to, the second moving body, regarded as stationary for this purpose. Thus, let Fig. 18 represent a tank so mounted that it may move horizontally to the right with a uniform absolute velocity of u feet per second; and let water issue from the various openings as indicated. Theoretically, the following absolute and relative velocities will result:

Orifice	Relative Velocity (to tank)	Absolute Velocity (referred to the earth)	θ	$\cos \theta$
A	$V = \sqrt{2gh}$	$v = V + u$	0°	1
B	$V =$ "	$v = V - u$	180°	-1
C	$V =$ "	$v = \sqrt{V^2 + u^2}$	90°	0
D	$V =$ "	$v = \sqrt{V^2 + u^2 + 2Vu \cos \theta}$	θ	$\cos \theta$ (positive)
E	$V =$ "	$v = \sqrt{V^2 + u^2 + 2Vu \cos \theta}$	θ	$\cos \theta$ (negative)

The expression for absolute velocity from orifice D or E may be regarded as a general formula, and the formulæ for the other cases may be simply derived from it by assigning the proper values to θ. These considerations of absolute and relative velocities are of great importance in determining the dynamic pressures produced by a stream of water on the moving vanes or blades of water-motors. For example, consider Fig. 19, which represents a revolving wheel having an orifice from which water issues horizontally with the relative velocity V (velocity relative to wheel), while the orifice itself is moving horizontally with an absolute velocity u (velocity relative to the ground); then, from what has preceded,

$$v = \sqrt{V^2 + u^2 + 2Vu\cos\theta} \quad (29)$$

is the absolute velocity of the water as it leaves the wheel (velocity with respect to the ground).

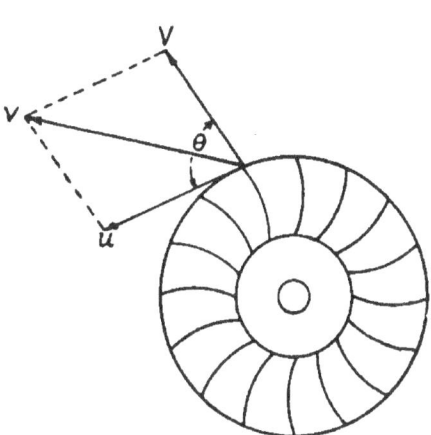

Fig. 19. Velocity of Stream Leaving or Striking Revolving Vane.

In all cases, then, *the absolute velocity of a stream of water striking or leaving a moving surface is represented in magnitude and direction by the diagonal of a parallelogram of which one side is the velocity of the stream relative to the moving surface, and the other side is the absolute velocity of that surface (with reference to the ground); i. e., it is the resultant of these two velocities.*

If the directions of the component velocities lie in the same straight line, $\theta = 0°$ or $180°$; and, applying Equation 29, we derive the special formulae:

$$v = V + u; \text{ or, } v = V - u \quad (29a)$$

SURFACES MOVING IN A STRAIGHT LINE

34. Dynamic Pressure on Moving Surfaces. When a stream of water impinges upon a moving surface, the conditions are essentially different from those just discussed for surfaces at rest. Because the surface is continually moving away from the stream, two important results follow—the stream does not strike the surface with its full or absolute velocity, and the quantity of water reaching the surface per second is less than the stream discharge.

35. Case I. Jet Striking a Moving Flat Vane Normally. Let a jet (Fig. 20) whose absolute velocity is v, and cross-section a, impinge normally upon a smooth surface which is itself moving with a uniform absolute velocity u in the same direction as the jet. The *relative* velocity of the jet, or the velocity with which it strikes the surface, is $v-u$; the weight of water *leaving the orifice* per second is

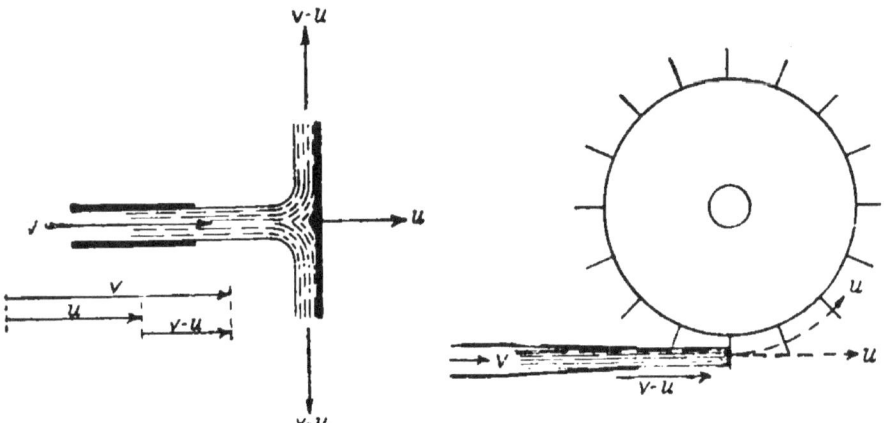

Fig. 20. Jet Striking a Moving Flat Vane Normally.   Fig. 21. Jet Striking Flat Radial Vanes of a Revolving Wheel.

$W = wav$; the weight of water *striking the surface* per second is $wa(v-u)$, if w represents the weight of a cubic unit of water; accordingly, the dynamic pressure exerted upon the surface, in the direction of motion, is:

$$P = wa(v-u)\frac{(v-u)}{g} = \frac{wa}{g}(v-u)^2, \quad \ldots (30)$$

which is equivalent to considering the surface stationary, and the stream moving with an absolute velocity of $(v-u)$ feet per second.

36. *Work Done upon (or Given Up to) the Moving Body per Second.* The work done in one second by the force P (Force × Distance) is:

$$\text{Work} = Pu = \frac{wa(v-u)^2 u}{g} \quad \ldots (31)$$

The work is zero if $u = v$; or $u = 0$; and it is a maximum, and equal to:

$$\text{Work (Max.)} = \frac{4}{27}\frac{wav^3}{g} = \frac{8}{27}W\frac{v^2}{2g} \ldots (32)$$

when $u = \tfrac{1}{3}v$.

37. *Efficiency.* Since the theoretic energy of the impinging jet is $W\frac{v^2}{2g}$, the efficiency in the case just considered is $\tfrac{8}{27}$, or about

30 per cent. It is evident, however, that no practical motor could be constructed on such a plan.

CASE II. This represents a wheel (Fig. 21) provided with many flat radial vanes against which, in rapid succession, a jet of water impinges. The resultant action of the jet in this case is not precisely the same as in the preceding example; but if we assume that the jet impinges normally on the vanes, and that, as the vanes come in rapid succession under the influence of the jet, and several vanes are more or less under action at the same time, the quantity of water impinging is the same as the nozzle discharge ($W = wav$); also, that the vanes move away from the jet in the direction of the latter while under impact, then we obtain for the *approximate* value of the dynamic pressure, if u represents the linear absolute velocity of the vanes at the center of impact:

$$P = W\frac{v-u}{g} = wav\frac{v-u}{g} \quad \ldots \ldots (33)$$

38. *Work Done upon (or Given Up to) the Wheel per Second.*

$$\text{Work} = Pu = \frac{wav}{g}(v-u)u \quad \ldots \ldots (34)$$

The work is zero if $u = v$, or $u = 0$; and it is a maximum and equal to:

$$\text{Work (Max.)} = \tfrac{1}{4}\frac{wav^3}{g} = \tfrac{1}{2}W\frac{v^2}{2g} \quad \ldots (35)$$

when $u = \tfrac{1}{2}v$.

39. *Efficiency.* Since the jet has a theoretic energy of $W\frac{v^2}{2g}$ foot-pounds, it is seen that the highest efficiency that can theoretically be obtained by means of a jet impinging upon rotating flat vanes is 50 per cent.

The preceding analysis applies more directly to the case of a series of flat vanes moving in a straight line, as indicated in Fig. 20, and coming in rapid succession under the influence of the jet. A motor constructed on this plan is, however, impracticable.

40. CASE III. JET STRIKING A MOVING CURVED VANE TANGENTIALLY. Fig. 22 represents a case in which the jet, with an absolute velocity v, impinges tangentially upon a vane which moves in the same direction with the uniform absolute velocity u. The velocity of the stream relative to the surface is $v - u$; and the dynamic pressure is the same as though the surface were at rest, and the stream

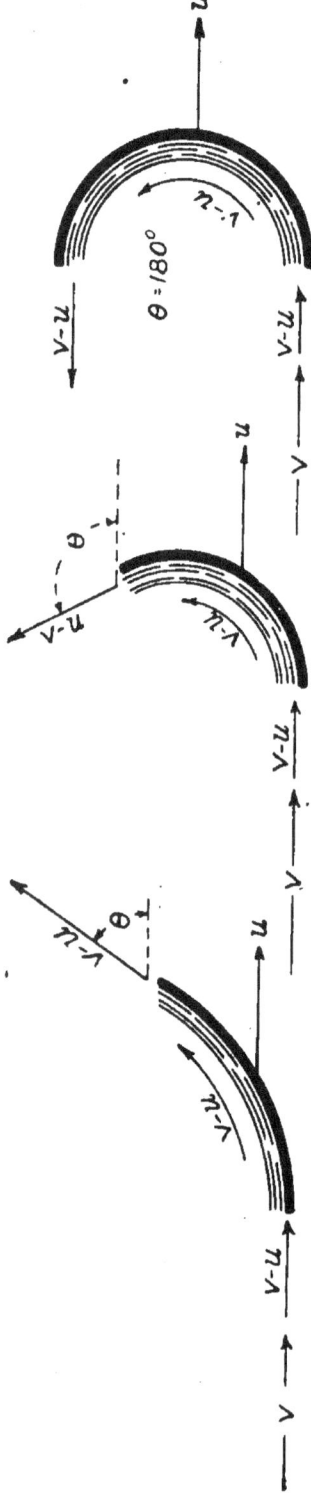

Fig. 22. Jet striking Moving Curved Vane Tangentially.

moving and impinging with the absolute velocity $v-u$. Hence, for the dynamic pressure in the direction of the jet, we may use Equation 26, substituting $v-u$ for v; so that,

$$P = (1-\cos\theta)W\frac{v-u}{g} \quad \ldots \quad (36)$$

While the dynamic pressure may be exerted with different intensities upon different parts of the vane, the total value, in the direction of motion, is that indicated by Equation 36.

41. Work Done. If a is the area of the cross-section of the jet, the weight of water issuing from the nozzle per second is $W = wav$; the weight striking the vane is $wa(v-u)$; and therefore the work is:

$$\text{Work} = Pu = (1-\cos\theta)\frac{wa}{g}(v-u)^2 u$$
$$\ldots \ldots \ldots \ldots \ldots (37)$$

The work is zero when $v = u$, and when $u = 0$; also when $\theta = 0°$; and it is a maximum, and equal to:

$$\text{Work (Max.)} = \frac{4}{27}(1-\cos\theta)wa\frac{v^3}{g} = \frac{8}{27}(1-\cos\theta)W\frac{v^2}{2g} \ldots \ldots (38)$$

when $u = \frac{1}{3}v$.

42. Efficiency. Since the theoretic energy of the impinging jet is $W\frac{v^2}{2g}$, the efficiency is:

$$e = \frac{8}{27}(1-\cos\theta) \ldots \ldots \ldots (39)$$

If $\theta = 0°$, work $= 0$, and $e = 0$; in this case the vane is a flat surface whose plane is in the direction of the stream, which therefore glides over the surface without doing work.

If $\theta = 90°$, the water leaves the vane at right angles to the direction of motion, and the maximum work, from Equation 38, is:

$$\text{Work (Max.)} = \tfrac{8}{27} W \frac{v^2}{2g} \quad \ldots \ldots (40)$$

and the efficiency is $\tfrac{8}{27}$, or about 30 per cent. (Compare with Equation 32.)

If $\theta = 180°$, the stream is completely reversed. In this case, (since $\cos 180° = -1$),

$$\text{Work (Max.)} = \tfrac{16}{27} W \frac{v^2}{2g} \quad \ldots \ldots (41)$$

and the efficiency is $\tfrac{16}{27}$, or about 60 per cent.

43. CASE IV. If, instead of a simple curved vane, as in the preceding case, we consider a wheel with a large number of such vanes, as in Fig. 23, and assume the jet to impinge tangentially, and the vanes to move in the direction of the jet while under its influence, and also the quantity of water impinging to be equal to the nozzle discharge, by an analysis similar to that which has preceded, we obtain:

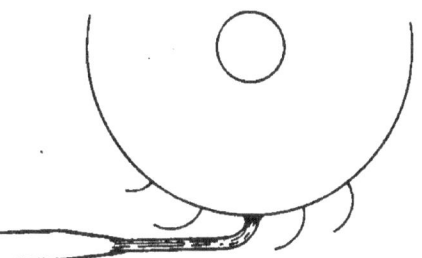

Fig. 23. Jet Striking Curved Vanes of a Revolving Wheel Tangentially.

44. The work is:

$$\text{Work} = (1 - \cos \theta) W \frac{(v - u)u}{g} \quad \ldots (41a)$$

This is zero when $u = 0$, or when $u = v$; also when $\theta = 0°$; and it is a maximum, and equal to:

$$\text{Work (Max.)} = \tfrac{1}{2}(1 - \cos \theta) W \frac{v^2}{2g} \quad \ldots (42)$$

when $u = \tfrac{1}{2} v$.

45. *Efficiency.* The efficiency is:

$$e = \tfrac{1}{2}(1 - \cos \theta) \quad \ldots \ldots (43)$$

When $\theta = 0°$, the stream merely glides along the surface without doing work, and $e = 0$.

When $\theta = 90°$, the jet is deflected normally to the direction of motion, and,

$$\text{Work (Max.)} = \tfrac{1}{2} W \frac{v^2}{2g} \quad \ldots \ldots (44)$$

and efficiency is $e = \tfrac{1}{2}$, or 50 per cent, as for radial flat vanes.

When $\theta = 180°$, the stream is completely reversed, and

$$\text{Work (Max.)} = W\frac{v^2}{2g} \quad \ldots \ldots \ldots \ldots (45)$$

in which case the efficiency is $e = 1$, or 100 per cent. The preceding analysis applies more directly to the case of a series of curved vanes moving in a straight line parallel to the jet, and coming in rapid succession under its influence. Such a motor is evidently impracticable.

46. In applying these considerations to water wheels, we must bear in mind that losses due to impact and friction have not been considered. The conclusions are therefore, to that extent, theoretic; but they represent limiting values which may be approached more and more closely, as the frictional and other resistances are reduced by means of correct design and construction. In the case of the conditions represented by Equation 45, since the efficiency is theoretically 100 per cent, it is clear that all the energy of the jet has been given up to the wheel, which would indicate that the absolute velocity of the water leaving the vanes must be zero; for if the water thus leaving has any absolute velocity, it still possesses some energy after passing clear of the wheel, which represents a portion of the original energy of the jet which has not been imparted to the wheel; the efficiency then could not be 100 per cent. This conclusion may be readily reached from the preceding analysis; for, since the best *absolute* velocity of the vane is $\frac{1}{2}v$, the water upon its surface has the *relative* velocity $v - \frac{1}{2}v = \frac{1}{2}v$, which is the same as the velocity of the vane, but in the *opposite direction;* then, if $\theta = 180°$, as in the case under discussion, the *absolute velocity of the water* as it leaves the vane, is $\frac{1}{2}v - \frac{1}{2}v = 0$.

While the above discussion shows that for maximum efficiency the velocity of the vanes should be one-half the velocity of the jet, the efficiency is not much lowered by slight variations of the vane velocity above or below the value indicated. It is also clear that to thus realize the full energy of the stream, we suppose the jet to both enter and leave the vanes in a direction tangential to the circumference, and a complete reversal is effected. It will be shown in a subsequent article that certain practical considerations render it impossible to fully realize these theoretic conditions.

47. If the vanes are plane radial surfaces, as in Fig. 21, the water passes from the wheel normally to the circumference, and

the highest obtainable efficiency is (theoretically) 50 per cent (Equation 35). In this case the water leaving the wheel still possesses absolute velocity to the extent of $\frac{v}{2}$, the component of which, in the direction of motion of the vanes, is $\frac{1}{2}v$; this represents a dynamic pressure of $W\frac{\frac{1}{2}v}{g}$ pounds in that direction, or $W\frac{\frac{1}{2}v}{g} \times \frac{1}{2}v \;(= P \times u) = \frac{1}{2}W\frac{v^2}{2g}$ foot-pounds of work; that is, one-half of the original energy of the jet is carried away by the escaping water, and is thus lost to the wheel. Or, an absolute velocity of $\frac{v}{2}$ represents kinetic energy to the amount of $\frac{W(\frac{v}{2})^2}{2g} = \frac{1}{2}W\frac{v^2}{2g}$. Equation 58 shows even more clearly that in order to realize the full theoretic energy of the stream, the absolute velocity of the departing water ($c_1 = \frac{v}{2}$ for this case) must be zero.

48. Case V. General. In the usual case the direction of motion of the vane is not the same as that of the jet. In Fig. 24, let the arrow marked v represent the direction of the jet as it impinges on the vane with an absolute velocity v; and let the arrow marked u represent the direction of motion of the vane, as well as its absolute velocity. While this case can be analyzed and solved in a manner similar to that employed in the preceding cases, it will be well here to adopt another procedure illustrating an important and useful principle:

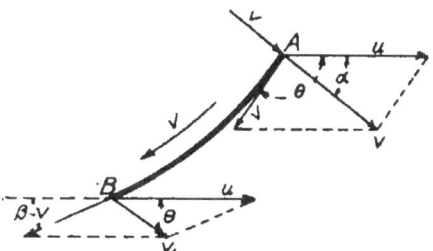

Fig. 24. General Case of a Jet Impinging on a Moving Vane.

The difference between the components of the absolute impulses of the entering and departing streams, in the direction of motion, is the resultant dynamic pressure in that direction.

49. *Dynamic Pressure in a Given Direction.* The absolute velocity of entry v being known, it remains to determine the absolute velocity of exit, v_1. By means of the principle enunciated in Article 33, we first find the relative velocity V with which the jet strikes the surface at A, by drawing to scale the lines v and u (both known) and completing the parallelogram. V then represents, in intensity and direction, the relative velocity of the stream at A. The stream passes over the surface, and leaves it at B with this same relative velocity, if not retarded by friction or shock. Now, by the principle just referred to and used for the point A, the absolute velocity of the stream as it leaves the vane at B may be determined. Draw u and V, and complete the parallelogram; v_1 then represents the absolute velocity of the escaping water at B.

The absolute impulse of the stream before striking the vane at A is $W\dfrac{v}{g}$; its component in the direction of motion is $W\dfrac{v}{g}\cos a$. The absolute impulse of the stream as it leaves the vane at B is $W\dfrac{v_1}{g}$; its component in the direction of motion is $W\dfrac{v_1}{g}\cos \theta$. Hence the dynamic pressure in the direction of motion is:

$$P = W\frac{v\cos a - v_1 \cos \theta}{g} \quad \ldots \ldots (46)$$

This is a general formula for the dynamic pressure in any given direction exerted by a jet of water upon a vane moving in a direction parallel to a straight line, if a and θ be the angles between that direction and the directions of v and v_1.

If the surface is at rest, $v = v_1$, and Equation 46 becomes $P = (\cos a - \cos \theta) W\dfrac{v}{g}$, which is Equation 27.

50. Usually, in the case represented by Fig. 24, the angles a and β are known, or assumed, and θ is unknown; it therefore becomes desirable to express the angle θ in other and known terms. By taking the components of the velocities at B in the direction of motion, it is evident that $v_1 \cos \theta = u - V \cos \beta$; if this value be substituted in Equation 46, there will result:

$$P = W\frac{v\cos a - u + V \cos \beta}{g}, \quad \ldots \ldots (47)$$

in which,

$$V^2 = u^2 + v^2 - 2uv \cos a \text{ (Trigonometry, from the triangle } A\, u\, v) \ldots (47a)$$

51. *Curvature of Vane at Entrance.* In order that the stream may strike the vane without shock, the curve of the vane at A should be tangent to the direction of V. It therefore becomes important to express the angle ϕ in known terms. From either triangle at A,

making use of the trigonometric principle that the sides of any plane triangle are proportional to the sines of their opposite angles, we obtain:

$$\frac{\sin(\phi - \alpha)}{\sin \phi} = \frac{u}{v} \quad \ldots \ldots \ldots (48)$$

which may be reduced, by known trigonometric relations, to:

$$\cot \phi = \cot \alpha - \frac{u}{v \sin \alpha} \quad \ldots \ldots (48a)$$

Equation 48a determines the angle ϕ, when u, v, and the angle α are known; and this fixes the proper curvature of the vane at the point A.

Example 14. In Fig. 24, let $u = 70.71$, $v = 100$, $\alpha = 45°$, and $\beta = 30°$. What is the dynamic pressure on the vane in the direction of motion, when 1 cubic foot of water strikes the vane per second? What should be the value of the angle ϕ in order that no loss by impact may occur?

From Equation 47a:

$$V = \sqrt{70.71^2 + 100^2 - 2 \times 70.71 \times 100 \times .707} = 70.71 \text{ feet per second.}$$

From Equation 47:

$$P = 62.5 \frac{100 \times .707 - 70.71 + 70.71 \times .866}{32.2} = 1,356 \text{ pounds.}$$

From Equation 48a:

$$\cot \phi = 1 - \frac{70.71}{100 \times .707} = 0 ; \therefore \phi = 90°$$

REVOLVING SURFACES

52. Case VI. In the case of water motors, the vanes upon which the jet impinges revolve about an axis. The motion of every point on the vane is therefore circular; hence, at any instant, the direction of motion of any point is tangent to the circumference drawn through, or it is normal to the radius drawn to, that point. At any point, therefore, that portion of the dynamic pressure which is effective in producing motion is its component in the direction of motion of that point. Fig. 25 illustrates two cases of wheels with vertical axes, the vanes revolving in horizontal planes. In the one case (B), the water, after impinging, passes outward, or away from the axis; in the other (a), the stream passes inward, or toward the axis. The following analysis, however, is general, and therefore applies to both types. As heretofore, v and v represent the absolute, and V and V_1 the relative velocities of the entering and departing streams; u and u_1 (drawn normal to the radii r and r_1) represent the absolute velocities and directions of motion of the points A and B on the vane; the angles to be used in the analysis are sufficiently clear from the diagram, in view of what has preceded. Constructing the two parallelograms in the usual manner, there is obtained, at

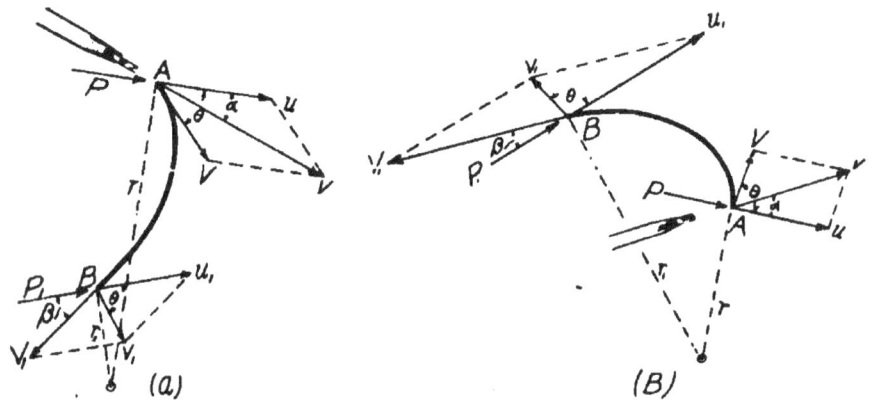

Fig. 25. Wheels with Vertical Axes, the Vanes Revolving in Horizontal Planes.

the point A, V as the relative velocity of the entering stream; and at the point B, v_1 as the absolute velocity of the departing stream. For the parallelogram at B, however, the value of V_1 must first be computed by means of Equation 54.

53. *Components of Pressures in Direction of Motion.* The total dynamic pressure exerted in the direction of motion will depend upon the impulses of the entering and the departing streams. The absolute impulse of the water on entering is $W\frac{v}{g}$; and that of the water on leaving is $W\frac{v_1}{g}$. The components of these in the directions of the motion of the vane at entrance and departure, are respectively:

$$P = W\frac{v \cos \alpha}{g} \; ; \text{ and } P_1 = W\frac{v_1 \cos \theta}{g} \quad \ldots \ldots (\mathbf{49})$$

Since their directions are not parallel, and the velocities of the points A and B are not equal, their difference cannot be taken to give the resultant dynamic pressure, as was done in Case V, which represented motion in a straight line; but this resultant pressure is not important. The two expressions in Equation 49, however, are useful in an analysis of the work that can be delivered by the vane.

54. Useful Formulæ. Since in any rotating body the linear velocities of points are directly proportional to their distances from the axis of rotation,

$$\frac{r}{r_1} = \frac{u}{u_1} \quad \ldots \ldots \ldots \ldots (50)$$

The relative velocities V and V_1 are connected with the velocities of rotation by the following simple relation:

$$V_1^2 - V^2 = u_1^2 - u^2 \quad \ldots \ldots \ldots (51)$$

Ordinarily, for a revolving vane, the data given or assumed will be the angles a, ϕ, and β; the radii r and r_1; the absolute velocity of the jet, v; the number of revolutions per second, n; and the weight of water delivered to the vane per second, W. Then,

$$u = 2\pi r n; \text{ and } u_1 = 2\pi r_1 n, \quad \ldots \ldots (52)$$

from which u and u_1 may be determined.

In the triangle Auv (sides are proportional to sines of opposite angles),

$$V = \frac{r \sin \alpha}{\sin \phi}, \quad \ldots \ldots \ldots \ldots (53)$$

which determines the relative velocity of entrance, V.

From Equation 51:

$$V_1 = \sqrt{u_1^2 - u^2 + V^2}, \quad \ldots \ldots \ldots (54)$$

which gives the value of the relative velocity of exit, V_1. Finally, taking the components of the velocities at B in the direction of motion of that point, there results:

$$r_1 \cos \theta = u_1 - V_1 \cos \beta \quad \ldots \ldots \ldots (55)$$

From the above equations, the numerical values of P and P_1 of Equation 49 can be fully determined.

Example 15. In Fig. 25; suppose $r = 2$ ft.; $r_1 = 3$ ft.; $\alpha = 45°$; $\phi = 90°$; $v = 100$ ft. per second; $n = 6$ revolutions per second. Compute the velocities u, u_1, V, and V_1.

From Equation 52:

$$u = 2 \times 3.1416 \times 2 \times 6 = 75.4 \text{ feet per second.}$$
$$u_1 = \frac{3}{2} u = 113.1 \text{ " " "}$$

From Equation 53:

$$V = \frac{100 \times .707}{1} = 70.71 \text{ feet per second.}$$

From Equation 54:

$$V_1 = \sqrt{113.1^2 - 75.4^2 + 70.71^2} = 110 \text{ feet per second.}$$

55. Work Derived from Revolving Vanes. In the discussion of "Work" and "Efficiency" under Cases IV and V, it was assumed that all points of the vane move with the same velocity; and in Case IV, that the stream enters upon it in the same direction as that of motion, or that $a = 0$. Considering the general case just discussed, it may be said that the work of a series of vanes arranged around a wheel may be regarded as that due to the absolute impulse of the entering stream in the direction of motion of the point of entrance, minus that due to the absolute impulse of the departing stream in the direction of motion of the point of exit; or,

$$\text{Work} = Pu - P_1 u_1 \quad \ldots \ldots \ldots \ldots (56)$$

in which P and P_1 are the components of the dynamic pressures due to the absolute impulses at A and B, in the directions of motion of the points A and B, respectively, as shown in Fig. 25 and Equation 49. Using the values of Equation 49, in Equation 56, there results:

$$\text{Work} = W \frac{uv \cos a - u_1 v_1 \cos \theta}{g} \ldots \ldots (57)$$

This is a perfectly general formula, applicable to the work of all wheels with outward or inward flow. It shows that the useful work consists of two parts—one due to the entering, and the other to the departing stream.

Another very simple general expression for the work of a series of revolving vanes may be deduced as follows: The total absolute energy of the entering stream is $W \frac{v^2}{2g}$; the total absolute energy of the departing stream is $W \frac{v_1^2}{2g}$; hence, neglecting friction and other resistances, the difference represents the energy imparted to, or taken up by, the wheel from the stream; that is:

$$\text{Work} = W \frac{v^2 - v_1^2}{2g} \ldots \ldots \ldots \ldots (58)$$

which is a useful formula of wide applicability. From Equation 58, the efficiency is:

$$e = \frac{v^2 - v_1^2}{v^2} = 1 - \left(\frac{v_1}{v}\right)^2 \ldots \ldots (59)$$

Example 16. As a numerical example, consider the case of the outward-flow horizontal wheel driven by a jet from a fixed nozzle, shown in Fig. 26.

Let $r = 2$ feet;
 $r_1 = 3$ feet;
 $\alpha = 45°$ (approach angle);
 $\phi = 90°$ (entrance angle);
 $\beta = 15°$ (exit angle);
 $v = 100$ feet per second;
 $q = 2.2$ cubic feet per second;
 $n = 337.5$ revolutions per minute.

It is required to find the useful work of the wheel, and its efficiency.

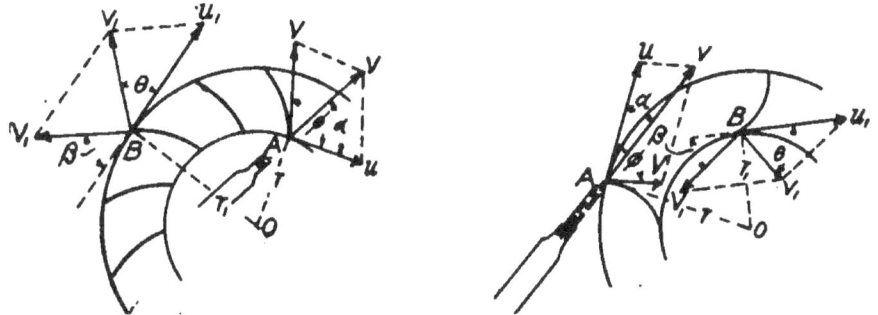

Fig. 26. Horizontal Wheels Driven by Jets from Fixed Nozzles.

From Equation 52:
$$u = 2\pi r n = 2 \times 3.1416 \times 2 \times \frac{337.5}{60} = 70.71 \text{ feet per second;}$$

and, from Equation 50:
$$u_1 = \frac{r_1}{r} u = \frac{3}{2} \times 70.71 = 106.06 \text{ feet per second.}$$

From Equation 53:
$$V = \frac{v \sin \alpha}{\sin \phi} = \frac{100 \times \sin 45°}{\sin 90°} = 100 \times 0.7071 = 70.71 \text{ feet per second.}$$

From Equation 54:
$$V_1 = \sqrt{u_1^2 - u^2 + V^2} = \sqrt{(106.06)^2 - (70.71)^2 + (70.71)^2} = 106.06 \text{ feet per second.}$$

From Equation 55:
$$r_1 \cos \theta = u_1 - V_1 \cos \beta = 106.06 - 106.06 \times \cos 15° = 3.61$$

Then, from Equation 57:
$$\text{Work} = 2.2 \times 62.5 \frac{70.71 \times 100 \times 0.707 - 106.06 \times 3.61}{32.2} = 19,712 \text{ ft.-lbs. per second.}$$

$$\frac{19,712}{550} = 35.8 \text{ horse-power.}$$

The theoretic energy of the jet is:

$$W \frac{v^2}{2g} = 2.2 \times 62.5 \frac{(100)^2}{64.4} = 21{,}380 \text{ ft.-lbs. per second.}$$

$$\frac{21{,}380}{550} = 38.9 \text{ horse-power.}$$

Therefore the efficiency of the wheel is:

$$e = \frac{19{,}712}{21{,}380}, \text{ or } \frac{35.8}{38.9} = 92.2 \text{ per cent.}$$

This would seem to indicate a very high efficiency; but it must be borne in mind that losses in friction, shock, etc., have not been considered in the preceding analyses. The effect of such resistances will be to reduce the computed efficiency.

Example 17. In the above example, assume the same data, except that $\beta = 30°$.

The values of u, u_1, V, and V_1 are not altered.

$$v_1 \cos \theta = 106.06 - 91.85 = 14.21$$

and,

$$\text{Work} = 14{,}910 \text{ ft.-lbs. per second,}$$
$$= 27.2 \text{ horse-power.}$$
$$\text{Efficiency} = 70 \text{ per cent.}$$

In both of the above examples the work and efficiency may be simply computed from Equations 58 and 59, after the value of v_1 has been determined. From Fig. 24, parallelogram at B, since u_1 and V_1 are equal in the above examples, it follows that $\theta = \frac{1}{2}(180 - \beta)$; therefore, from Equation 55:

$$v_1 = \frac{u_1 - V_1 \cos \beta}{\cos \theta} = \frac{u_1 - V_1 \cos \beta}{\sin \frac{1}{2} \beta}$$

$$= \frac{106.06\,(1 - .966)}{.131} = 27.52 \text{ (for example 16);}$$

and,

$$v_1 = 106.06 \frac{(1 - .866)}{.259} = 54.87 \text{ (for example 17).}$$

Substituting numerical values in Equations 58 and 59, the same results for the work and efficiency will be found as computed before.

HYDRAULIC MOTORS

56. Definition. A *hydraulic motor* may be defined as a machine in which the energy stored in water is utilized to produce motion and thus perform work. The energy of water, as was explained in Article 6, may exist in the form of gravity, of pressure, or of velocity; of these, gravity and pressure are not essentially or fundamentally separate and distinct phenomena, but rather the result of considering the weight of the water from different points of view. In general, then, it may be said that a hydraulic motor is an apparatus (usually a wheel) which is caused to move (usually rotate) by reason of a weight of water falling from a higher to a lower level, or because of the dynamic pressure induced by a change of direction, or of velocity, or both, in a moving stream. The dynamic pressure may be due to impulse, or reaction, or both. Many wheels are actuated by a combination in varying proportion of the above agencies, which are but manifestations of the energy existing in the water.

57. General Requirements for High Efficiency. The efficiency of any motor should, if possible, be independent of the quantity of water supplied to it; or, if the efficiency does vary with the supply, it should, when possible, be greatest in time of low water. It has already been shown that when W pounds of water fall through a height of h feet, or are delivered with a velocity of v feet per second, the theoretic energy in foot-pounds per second is:

$$K = Wh; \text{ or } K = W\frac{v^2}{2g}.$$

58. The actual work performed, or that may be performed, per second is equal to the theoretic energy, *minus* all the losses of energy. It is convenient to subdivide these losses into four general classes:

(a) Losses incidental to the conduction of the water from the supply to the motor, occasioned by friction and the various other resistances usually encountered, such as bends, changes of section, passages through orifices or other controlling devices which are not essentially parts of the apparatus itself, etc.;

(b) Losses in passage through the motor, which include friction, losses in eddies resulting from abrupt change in cross-section and improper entrance angle, and losses in passage through controlling devices which form part of the apparatus, etc.;

(c) The residual energy still possessed by the departing water flowing away with an absolute velocity v_1;

(d) Shaft and journal friction.

Sometimes the friction of the moving parts in the air or water is included, but will not here be considered.

59. Efficiency. Let Wh' represent the energy lost in conduction; Wh'', that lost in passage through the wheel; $W\frac{v_1^2}{2g}$, the energy still remaining in the departing water; and Wh''', the energy lost in shaft and journal friction; then,

$$k = W(h - h' - h'' - \frac{v_1^2}{2g} - h''')$$

represents the actual useful work per second that the wheel is capable of performing. Accordingly, if v is the velocity due to the head h, the efficiency is:

$$e = \frac{k}{K} = 1 - \frac{h'}{h} - \frac{h''}{h} - \left(\frac{v_1}{v}\right)^2 - \frac{h'''}{h}.$$

This formula, being very general, leads to the four following broad statements of the conditions requisite for high efficiency:

(1) The water must be conducted to the motor, and
(2) The water must pass through the motor, with the minimum loss of energy.
(3) The water must reach the tail-race level with the minimum absolute velocity consistent with practical considerations, such as the necessity for quick and proper clearance of water from the buckets, etc.
(4) The friction and other mechanical resistances of the moving parts must be reduced to a minimum.

60. This analysis, with the corresponding formulæ, compares the energy of the entire waterfall with the ultimate output of the machine. In estimating the power and efficiency yielded by the motor itself, *regarded as a user of water delivered to it with a definite amount of energy*, certain of the above losses should be omitted. Thus, losses in the conduction of the water to the motor cannot properly be charged against the motor; nor should losses in journal and shaft friction, which are outside and independent of the wheel regarded as a water user; in fact, the overcoming of journal and shaft friction is part of the work performed by the wheel, though it is not *useful* work. The energy in the departing water is properly chargeable to the wheel, since it is directly dependent upon the design or construction of the wheel. Therefore the hydraulic efficiency of the wheel may be stated thus:

$$e = 1 - \frac{h''}{h} - \left(\frac{v_1}{v}\right)^2; \quad \ldots \ldots \ldots \quad (60)$$

or, as popularly stated, for high efficiency "the water should enter the wheel without shock, and leave without velocity." When the actual power and efficiency of a water motor are practically measured as described in Articles 115 *et seq.*, the shaft and journal friction and air or water resistance are automatically included in the result. This explains why the results of actual tests of power and efficiency are always lower than the corresponding values computed from formulæ derived without consideration of such losses. It is therefore well to employ two terms, *hydraulic efficiency* and *actual efficiency*, in order to distinguish clearly between the two sets of conditions involved.

61. Classification. In the absence of a uniform or generally accepted classification, hydraulic motors may be divided into two general classes:

(*a*) *Water-wheels*, in which the water does not enter and actuate the wheel around the entire circumference.

(*b*) *Turbines*, in which the water enters and actuates the wheel around the entire circumference.

Each of these main divisions has several subdivisions.

WATER-WHEELS

62. Overshot Wheel. In this form of wheel, the water enters at the top and acts mainly by its weight; nevertheless, in most forms, an appreciable amount of kinetic energy is likewise imparted to the wheel. Fig. 27 shows a vertical section of such a wheel. The *buckets* are formed by vanes or partitions made in two parts— one part *a* in line with the radius of the wheel, the other part *b* inclined in a direction definitely determined by the design. The bottom of the bucket is formed by the rim or *sole-plate* F; the side pieces are made by two *cheeks* or *shrouds* E. The whole is bolted to arms assembled on the hub, and supported by the axle.

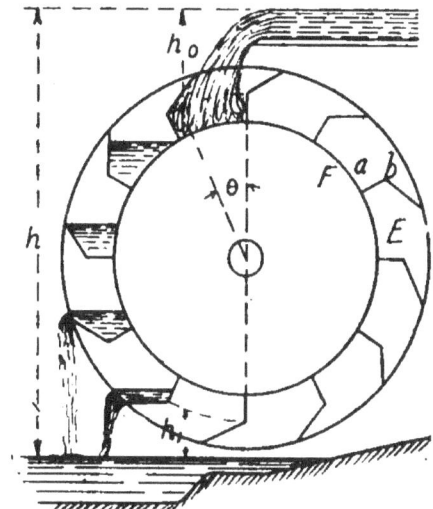

Fig. 27. Vertical Section of Overshot Wheel.

Let *h* be the total fall from the surface of the water in the head-race or flume to the surface of the water in the tail-race; and let W be the weight of water delivered to the wheel per second. The theoretic energy of the waterfall per second is Wh foot-pounds. The total fall *h* may be conveniently divided into three parts—namely,

h_0, the average head in filling the buckets; $h-h_0-h_1$, the average head of descent of the filled buckets; and h_1, that part of the head which remains between the empty buckets and the tail-race. The water strikes the buckets with a velocity v_0, approximately equal to $\sqrt{2gh_0}$; the buckets themselves are moving with a tangential velocity u approximately in the same direction as v_0; this occasions a loss of head in impact, h'' (Mechanics):

$$h'' = \frac{(v_0-u)^2}{2g}.$$

The water then descends through the average distance $h-h_0-h_1$, acting by its weight alone; finally it drops out of the buckets, and reaches the level of the tail-race with the absolute velocity v_1, which represents part of the original energy wasted. Accordingly, the efficiency of the wheel is:

$$e = 1 - \frac{h''}{h} - \frac{v_1^2}{2gh}.$$

Since the water leaving the buckets has a velocity u when commencing the descent through height h_1, its velocity at the level of the tail-race is:

$$v_1 = \sqrt{u^2 + 2gh_1}.$$

Substituting the values h'' and v_1 in the equation of efficiency above,

$$e = 1 - \frac{v_0^2 - 2v_0 u + 2u^2 + 2gh_1}{2gh},$$

and ascertaining by the usual procedure in such cases what value of u will render the efficiency e a maximum, it is readily found that:

$$u = \tfrac{1}{2}v_0; \quad \ldots \ldots \ldots \ldots (61)$$

that is, theoretically, the velocity of the wheel should be one-half that of the entering water for maximum efficiency. With this value of u, the hydraulic efficiency is:

$$e \text{ (Max.)} = 1 - \frac{1}{2}\frac{h_0}{h} - \frac{h_1}{h} \ldots \ldots (62)$$

and

$$\text{Work (Max.)} = Wh \times e = W(h - \frac{h_0}{2} - h_1) \ldots (63)$$

for the maximum efficiency and work of the overshot wheel. This equation teaches that one-half of the entrance drop h_0, and the whole of the exit drop h_1, are lost. Therefore, in order that the efficiency should be as high as possible, both h_0 and h_1 should be as small as practicable. The former requirement may be met by making the wheel of large diameter; but h_0 can never be zero, for in that case no water would enter the wheel; practically the size of wheel is usually such that θ equals 10 to 15 degrees. The fall h_1 is made small by giving to the buckets such a form that the water will be retained as long as possible, and by having as little clearance as practically advisable between the lowest point of the wheel and the tail-race level. In the design illustrated in Fig. 28, the buckets are deep in order to hold the water as long as possible; and moreover, they are shaped to conform to the direction of the entering water, thereby avoiding shock. Wheels of this description have been constructed 50 feet in diameter. In this case the power is taken from the axle of the small pinion, which is driven by a toothed ring attached to the circumference of the wheel. In other cases the power may be taken directly from the shaft of the water-wheel, through intermediate gearing, or by a crank-shaft.

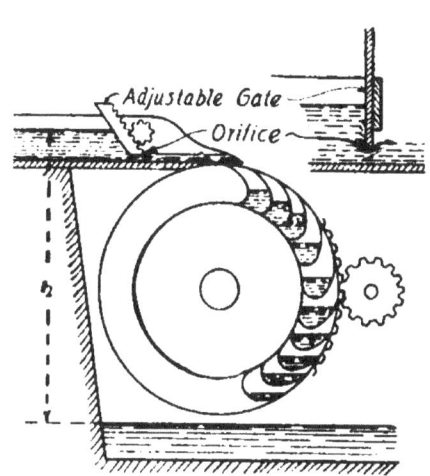

Fig. 28. Overshot Wheel with Deep Buckets to Hold Water as Long as Possible.

The method of regulating the supply of water to the wheel is also shown in the diagram. The theoretic advantageous velocity of the overshot wheel was shown to be $u = \frac{1}{2}v_0$; practically, this advantageous velocity is found to be about $u = 0.4 v_0$; and the efficiency of the wheel is high, ranging from 70 to 85 per cent, or over. One great advantage of the overshot wheel is that its efficiency is highest in times of drought, when the supply is low, for then the buckets are but partly filled, they do not begin to empty at as high a point above tail-water as when they are full; hence h_1 becomes small, with corresponding increase in efficiency. The main disadvantage of the overshot wheel lies in its size and its cost of construction. Moreover, its speed being slow (commonly from 3 to 6 feet peripheral velocity), it often requires the installation of somewhat complicated and expensive transmission gearing in order to drive machinery at a suitable speed; it is therefore best adapted to drive slow-moving machinery, usually with heads from 10 to 40 feet (though much larger heads have been used), and with a supply of from 100 to 350 gallons per second. A peripheral speed much greater than that commonly employed would result in a waste of water from the buckets due to centrifugal force.

The number of buckets and their depth are sometimes determined by formulæ, but they are largely matters of experience. If r is the radius of the wheel in feet, the number of buckets is usually $5r$ or $6r$, and their radial depth 10 to 15 inches. The width of the wheel parallel to the shaft is governed by the quantity of water actuating the wheel; it should preferably be so great that the buckets will not be quite full, thus reducing the fall h_1. If the tail-water level is constant, the lowest part of the wheel should be set just clear of that level; if it is variable, just sufficient clearance should be allowed to prevent interference and resistance in times of high water.

These precautions are necessary, for it is clear that the direction of motion of the buckets in the lowest portion of the wheel is opposite to the stream flow in the tail-race; and even slight submergence, therefore, will offer great additional resistance to its motion. This difficulty is sometimes obviated, when for any reason the wheel is to be submerged 4 or 5 inches (as by reason of variable tail-race level), by adopting a reverse-feed arrangement at the end of the supply channel, by which means the water is introduced on the back instead of on the front of the wheel, causing it to revolve in the opposite direction, so that the lower buckets move in the same direction as the tail-water. Such a wheel is often called a *back-pitch* or *back-shot* wheel.

For shallow streams of water with fairly constant depth, the supply channel is usually open-ended, as in Fig. 27; for deeper streams, or greater falls, the supply channel is provided with a sluice-gate or other regulating device, as in Fig. 28. Such a supply-regulating device is especially necessary in case of variable stream-flow.

Perhaps the largest overshot wheel in existence is that at Laxey, Isle of Man (Fig. 29), off the west coast of England. It is 72 feet 6 inches in diameter, and is said to yield 150 to 200 horse-power useful work, which consists in draining a mine 1,200 to 1,380 feet deep. The water for operating is conveyed to the wheel in an underground conduit, and is carried up the masonry tower by pressure, flowing over the top into the buckets of the wheel. Probably the

Fig. 29. Overshot Wheel at Laxey, Isle of Man.
Diameter of wheel, 72 ft. 6 in. Water carried up masonry tower by pressure, then flowing into buckets of the wheel.

largest wheel of this type in the United States was erected at Troy, N. Y., with a diameter of 62 feet and a width of 22 feet, developing 550 horse-power.

63. **Breast Wheel.** This type of wheel is designed to receive the water on one side, about or a little above the level of the hori-

zontal diameter; its lower portion, therefore, moves in the direction of the tail-water stream; for this reason the wheel may be *drowned*, or submerged, to a depth of 4 to 6 inches, which makes it suitable for use when both head-race and tail-race levels and supply are subject to variation. It is also evident from the manner of arranging the supply water, that this type is applicable only to small falls, from about 8 to 15 feet; for larger falls, the size of wheel would become impracticable. It is clear that the water acts both by impulse and by weight; therefore, to prevent the escape of the water before the buckets reach their lowest position, the lower quarter of the wheel is encased in a circular *breast* which encloses the buckets, thus practically compelling the water, or most of it, to remain therein until the lowest point is reached. In Fig. 30, water is conducted from the source in a channel or trough to and through an orifice A, which controls the supply to the wheel through regulation of the size of the orifice.

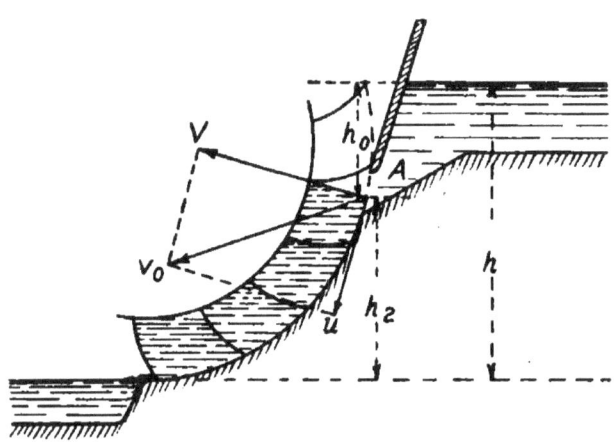

Fig. 30. Breast-Wheel with Supply Controlled through Size of Orifice.

In Fig. 31 the control of the supply is accomplished by means of a shuttle-gate arrangement which consists of a number of openings $J J$ in the inclined end of the trough, one or more of which may be closed by shifting the sliding gate B. The guide-pieces are for the purpose of causing the water to enter the buckets in a direction most favorable for good efficiency. With the arrangement indicated in Fig. 31, considering the way in which the water enters the buckets, and observing that the mouths of the buckets are practically covered by the extension of the guide-pieces, it is evident that vents or air-holes $F F$ in the sole-plate are necessary. Or the sole-plate may be dispensed with entirely, and the buckets formed of polygonal pockets, as $b\ a\ c$, in which the vents are naturally formed by the spaces left between the inner sides of consecutive buckets; these being at the top, the buckets may be completely filled with water.

Work and Efficiency. In Fig. 30, the water is admitted through the orifice A, under a head h_0; it therefore strikes the wheel with a velocity v_0, which is approximately equal to $\sqrt{2gh_0}$, and actually equal to $c_1\sqrt{2gh_0}$, where c_1 is the coefficient of velocity for the orifice at A. The water, being then confined between the vanes and the curved breast, acts by its weight alone through the distance h_2, which is approximately equal to $h - h_0$; finally it escapes at the level of the tail-race with the velocity u, or the velocity of the circumference of the wheel. The reasoning in the article on overshot wheels may be applied to this case, by making the fall h_1 equal to zero, and the resulting conclusions may be considered to apply approximately to the case of breast wheels. Accordingly, the following relations are approximately true:

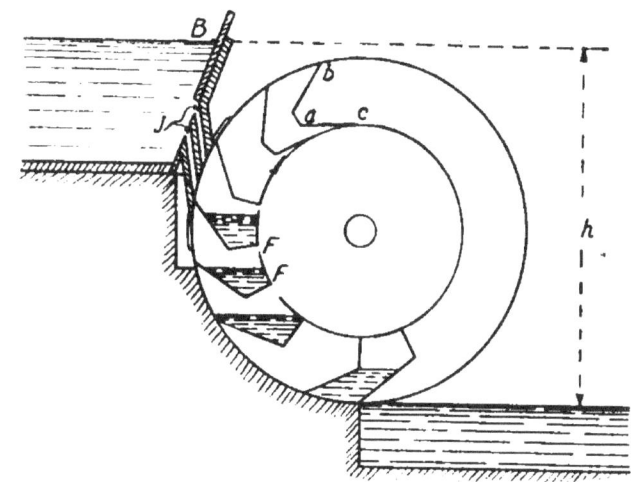

Fig. 31. Breast Wheel with Supply Controlled by Shuttle Gate.

The most advantageous theoretic velocity is

$$u = \tfrac{1}{2} v_0 = \tfrac{1}{2}\sqrt{2gh_0} \quad \ldots \ldots \ldots (64)$$

The maximum efficiency is theoretically:

$$e(\text{Max.}) = 1 - \tfrac{1}{2}\frac{h_0}{h} \quad \ldots \ldots \ldots (65)$$

The maximum work is theoretically:

$$\text{Work (Max.)} = W(h - \tfrac{1}{2}h_0) \quad \ldots \ldots \ldots (66)$$

Practically, the coefficient of velocity of the entrance orifice should be considered, as well as loss due to the clearance between wheel and breast, which will always exist; for any attempt to prevent this entirely by making the clearance less than about $\tfrac{3}{16}$ inch would result in a considerable increase in circumferential friction, and also, if the wheel is slightly off center, in repeated shocks. For these reasons the efficiency of the breast wheel is materially less than that of the overshot wheel, the usual values ranging from about 50 per cent for small wheels to about 75 per cent for large, well-designed wheels.

When the fall is not great, the wheel is sometimes designed to receive the supply water at a point appreciably below the horizontal diameter; in this case it is frequently termed a *side wheel*. Its efficiency is lower than that of the regular breast wheel. The best wheels of this type have been constructed with diameters ranging between 12 and 24 feet, running with circumferential velocities between 6 and 10 feet per second. They may be regarded as a type intermediate between the regular breast wheel and the undershot wheel. Breast wheels are sometimes provided with some simple automatic governing device controlled by the speed of the wheel, whereby the feed-water orifice is partially throttled when the speed of rotation exceeds a definite predetermined amount.

64. Undershot Wheel. The common undershot wheel is provided with plane radial vanes, and the wheel is so set that the water impinges on the lower vanes only, in an almost horizontal direction. In one sense, then, the undershot wheel may be regarded as a special kind of breast wheel, which is operated entirely by the impulse of the moving water. The formulæ developed for the case of breast wheels may therefore be applied approximately to the case of undershot wheels by changing h_0 to h, and v_0 to v; thus, for the most advantageous velocity of the wheel:

$$u = \tfrac{1}{2}v = \tfrac{1}{2}\sqrt{2gh}; \quad \ldots \ldots \ldots (67)$$

the maximum efficiency is:

$$e \text{ (Max.)} = \tfrac{1}{2}, \text{ or } 50 \text{ per cent}; \quad \ldots \ldots (68)$$

and the maximum work of the wheel is:

$$\text{Work (Max.)} = \tfrac{1}{2}Wh \quad \ldots \ldots \ldots \ldots (69)$$

Here, also, the coefficient of velocity of the water in passing through the orifice should properly be considered. In this type, as well as in the last, for reasons set forth in a preceding article, the maximum efficiency and maximum work are practically less than indicated in the foregoing formulæ; also, the most advantageous speed of the wheel is more nearly $u = .401\overline{2gh}$ than $.501\overline{2gh}$. In practice the efficiencies of such wheels are found to lie between 20 and 40 per cent. The lowest efficiencies are obtained from wheels placed in an unconfined current of water, such as a wheel attached to a barge anchored in a stream; and the higher efficiencies may be expected from well-constructed wheels, in which the actuating stream of water is properly confined, so that it cannot spread laterally.

Fig. 32 shows a simple type of radial-vane undershot wheel operating under a head of water. Here it is seen that the wheel is set in a circular channel constructed with a radius a trifle larger than that of the periphery of the wheel. The sliding gate for regulating the supply from the penstock is arranged at an angle of about 45°,

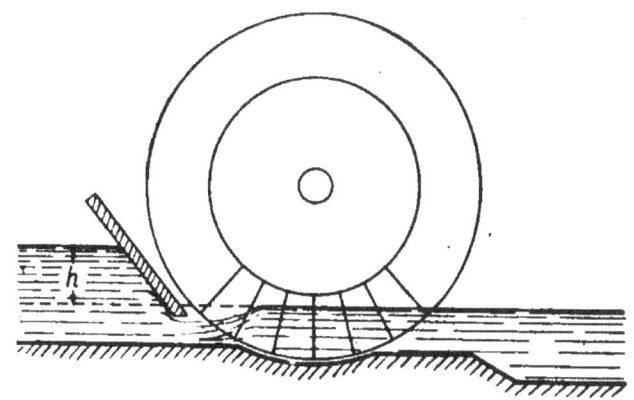

Fig. 32. Simple Type of Radial-Vane Undershot Wheel.

which enables its lower edge to be set close to the wheel rim. By this means the vanes are kept from contact with the moving water until they are almost vertical. The slight drop in the channel below the wheel compensates to some extent for the friction loss in passing the orifice of entry. The circular channel is succeeded by a gently inclined bed, so that the water maintains its uniform velocity after leaving the wheel, until, at a point well away from the wheel, the channel bed is given a sudden, steep inclination.

The depth of opening at the orifice usually varies from about 8 inches as a minimum, to about 20 inches in flood. The number of blades, N, is sometimes calculated from the empirical formula:

$$N = 4R,$$

in which R is the wheel radius. Then N and R will determine the spacing between the blades. In practice, this spacing may vary between 18 and 24 inches.

The undershot wheel is a relatively high-speed wheel; hence it may be made more compact than the types described before; its construction and installation are extremely simple, and, from these points of view, it is economical. But its efficiency is lower than that of the other types; it is suitable only for very simple installations, to drive machinery at relatively high speed, where an ample supply of water is available, under a low head.

65. **Poncelet Wheel.** In this wheel (Fig. 33), the vanes are curved in such a way that the water enters through the regulating orifice or opening without shock. Let v be the absolute velocity of the entering stream, and u the peripheral speed of the wheel. The stream, entering with the absolute velocity v, impinges tangentially on the smooth vanes, which are themselves moving in the same general direction with an absolute

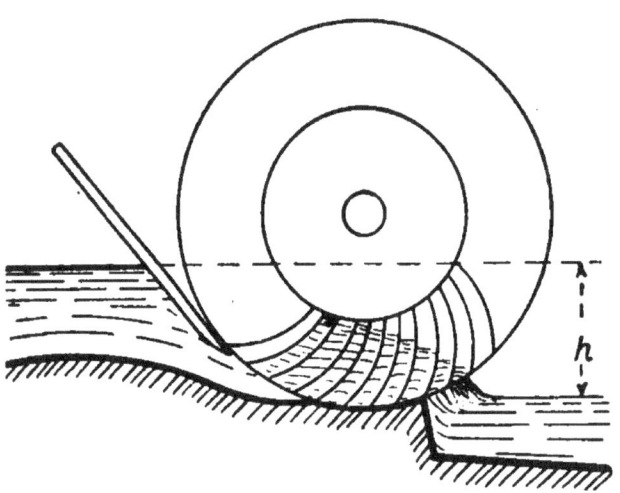

Fig. 33. Poncelet Wheel, Water Entering without Shock.

velocity u. The relative velocity of the water is therefore $v-u$; and it glides smoothly up the curved vane in the general direction of motion of the stream to a height corresponding to this velocity; when at its uppermost point, it is at rest relatively to the vane; it then falls, exerting pressure as it falls, gliding along the vane in the general direction opposite to the motion of the stream, attaining the velocity $v - u$ at the lowest point or extremity of the vane, and passing from the vane tangentially. Its dynamic pressure is therefore due to both impulse and reaction:

$$P = F + R = 2F = 2W\frac{v-u}{g};$$

the work of the wheel is $(P \times u)$;

$$\text{Work} = 2W\left(\frac{v-u}{g}\right)u;$$

and this is a maximum, and equal to:

$$\text{Work (Max.)} = W\frac{v^2}{2g} = Wh \quad \ldots \ldots \ldots \quad (70)$$

when $u = \frac{1}{2}v$.

Since the theoretic energy of the stream is $W\dfrac{v^2}{2g} = Wh$,

$$e \text{ (Max.)} = 1, \text{ or } 100 \text{ per cent} \dots \dots (71)$$

This follows from the fact that with the advantageous velocity $u = \frac{1}{2}v$, the absolute velocity of exit is zero; hence the stream "enters without impact, and departs without velocity."

The preceding analysis and the conclusions are theoretic, since they do not consider the various losses of head or energy which must take place. Practically, the efficiency lies between 65 and 75 per cent.

The curved form is given to the bed of the channel of approach, in order to direct the entering stream of water so as to avoid shock. The depth of the vane should be such that the entering water may run up its length (due to its relative velocity) without interference. The spacing of the blades usually ranges between 10 and 18 inches.

The Poncelet wheel, like other undershot wheels, has a relatively high speed; its efficiency is almost independent of the flow, and also of the speed, when a curved channel of approach is used. Moreover, this speed does not vary much, in spite of considerable variations of head. This form of wheel may be used to advantage with a head not exceeding about 6 feet, when the application of power does not call for a high velocity, as for pumping, grinding, etc.

66. In the foregoing cases, the analytical relations have been deduced largely by comparison and analogy, resulting in conclusions more or less approximately true. In each case, however, these relations may be developed quite independently, giving theoretically accurate results. For example, take the case of the breast wheel represented in Fig. 30. In the figure, let Av_0 and Au represent in intensity and direction the velocities, respectively, of the entering water and of the vanes, inclined to each other at an angle a. The dynamic pressure exerted by the water on the vanes, in the direction of motion, is:

$$P = W \cdot \dfrac{v_0 \cos a - u}{g};$$

and the work per second is:

$$K = W \dfrac{(v_0 \cos a - u)}{g} u. \dots \dots (72)$$

The work K, of the dynamic pressure alone, is a maximum, and equal to:

$$\text{Work (Max.)} = W \frac{v_0^2 \cos^2 \alpha}{4g} \quad \ldots \ldots (73)$$

when $u = \tfrac{1}{2} v_0 \cos \alpha$.

To this value of K must be added the term Wh_2, representing the work done by the weight of water in the buckets falling the distance h_2; this term is theoretically independent of the speed; accordingly,

$$\text{Total work (Max.)} = W\left(\frac{v_0^2 \cos^2 \alpha}{4g} + h_2\right); \ldots (74)$$

but $v_0 = c_1 \sqrt{2gh_0}$, where c_1 is the coefficient of velocity for the orifice at A. Therefore,

$$\text{Total work (Max.)} = W\left(\tfrac{1}{2} c_1^2 \cos^2 \alpha \cdot h_0 + h_2\right); \quad (75)$$

and the maximum hydraulic efficiency is:

$$e \text{ (Max.)} = \tfrac{1}{2} c_1^2 \cos^2 \alpha \frac{h_0}{h} + \frac{h_2}{h}. \ldots \ldots (76)$$

If, in these equations (73, 75, and 76), h_2 be replaced by its equal $h - h_0$, and if c_1 equals unity, and the angle α equals zero, there will result the approximate equations 64, 65, and 66, deduced in Article 63.

The angle α, however, cannot be zero; in fact it cannot practically be made less than about 10 degrees, for then little or no water would enter the wheel; it should, nevertheless, be as small as practicable, and is usually found between 10 and 25 degrees. The value of the coefficient c_1 is rendered large by well rounding the edges of the orifice; in this way c_1 may be made equal to .95 or even .98. In a manner similar to the above, formulæ for the other cases discussed may also be developed, with a greater degree of accuracy, theoretically considered. It is evident, however, that the approximate formulæ are sufficiently exact for most purposes, since the losses due to improper entry, foam, and leakage, cannot be algebraically expressed.

SPECIAL FORMS OF WHEELS

Water wheels in great variety have been in use from very early times, some of them operating with a fair degree of efficiency. A few of these forms will be very briefly described.

67. Sagebien Side Wheel. The buckets of this wheel (Fig. 34) are formed by flat vanes which are tangent to the horizontal cylinder O, whose axis is concentric with the shaft of the wheel. The depth of the bucket-ring is relatively large, and there is no sole-plate, each bucket forming a sort of vessel open on top and bottom. The wheel turns in a circular channel, prolonged upstream by a suitable iron casing, sometimes called a *swan's neck*. The side cheeks of the

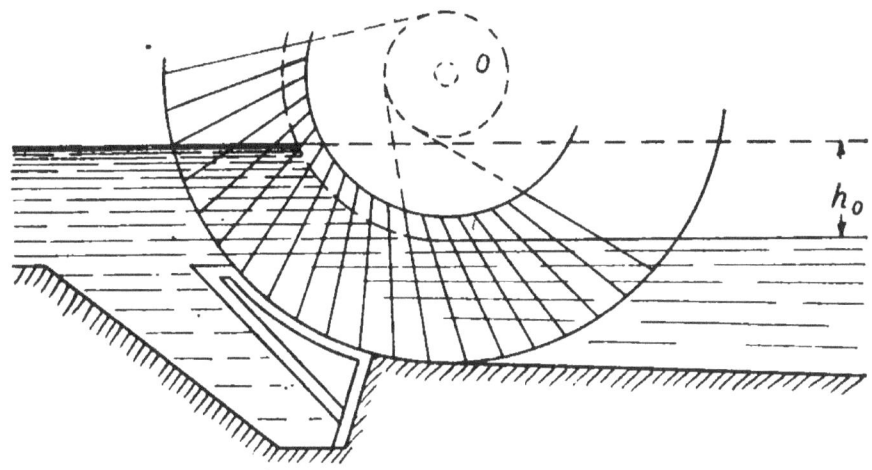

Fig. 34. Sagebien Side Wheel.

channel are continued downstream to the point where the wheel leaves the tail-race. There is very little work done by the water on the vanes beyond a point vertically below the center of the wheel. The inclination of the blades is not favorable for their easy emergence on the downstream side; but, as the speed of the wheel is rarely so great as 3 feet per second, being more usually between 1 and 2 feet, this resistance is small. The efficiency of this wheel, on account of its low speed (since resistances increase more or less rapidly with the speed), is very high, ranging from 80 to 90 per cent according to the height of the fall and the diameter, which varies between 20 and 40 feet, depending upon the fall available, the variability of the supply, and the fluctuation in the tail-race level. The number of revolutions per minute is often less than 1, and rarely exceeds $2\frac{1}{2}$. The penstock speed is usually 1 to 2 feet per second, and this is about the velocity with which the water enters the wheel. The spacing of the blades, measured on the outside of the wheel is about 15 inches. This type of wheel is used for small falls, from 2 to 9 feet, and is suitable for large flows. On account of its slow speed, it is adaptable only for installations where the machinery runs slowly and opposes uniform resistance to driving.

68. **Millot Wheel.** This is a form of breast wheel (Fig. 35) in which the breast is not needed. The supply channel divides into two branches, which pass around to the inner side of the wheel, so that the water enters at the inner circumference. This wheel is difficult to construct, and can be used only for small powers, since, by reason of the feed-water arrangement, the arms must be placed in the middle section of the wheel, instead of being fixed to the flanges; for this reason the breadth is limited to about 5 feet.

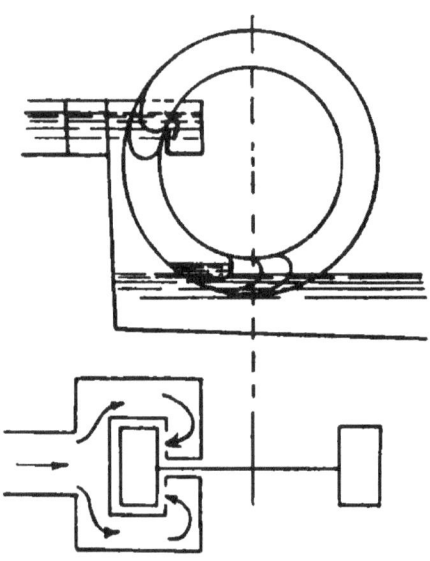

Fig. 35. Millot Wheel.

69. **Floating Wheel, or Current Wheel.** This type is simply an undershot wheel with flat, radial vanes, erected on a scow or barge intended to be anchored in a stream, or mounted on some suitable framework built up from the stream bed. The flat blades are attached to an inner circle, but are not enclosed in shrouds, so that the water has very free entry. As the barge rises and falls with the changes of the stream level, the depth of blade immersion is constant. The efficiency is theoretically a maximum, and equal to 50 per cent, when the peripheral speed of the wheel is one-half the velocity of the current; actually, it rarely reaches 40 per cent. When such a wheel is required to drive stationary machinery—that is, machinery so mounted that it does not follow the fluctuations in the surface level—some special device must be employed to insure the required condition of constant depth of paddle immersion. These wheels are extremely simple, but require to be of large size in order to develop even a moderate amount of power.

Wheels of this type have been used for operating dredges on the river Rhine, Germany; they have also been used to a limited extent, principally for irrigation purposes, in the western part of the

United States. One at Fayette Valley, Idaho, was said to be 28 feet in diameter, with 28 paddles, each 16 feet long and 2½ feet wide.

70. **Tympanium.** This is an ancient form of circular open-frame wheel (Fig. 36), fitted with radial partitions so directed as to point upward on the rising side of the wheel, and downward on the descending side. The wheel is mounted in such a way that its lowest parts are submerged to a convenient depth, and it may be turned by the impulse of the current impinging on radial vanes arranged around its circumference. The partitions scoop up a quantity of water, which, as the wheel revolves, runs back toward the axis, where it is discharged into a trough that conveys it away. A very evident disadvantage of this form of wheel is the fact that the water has to be

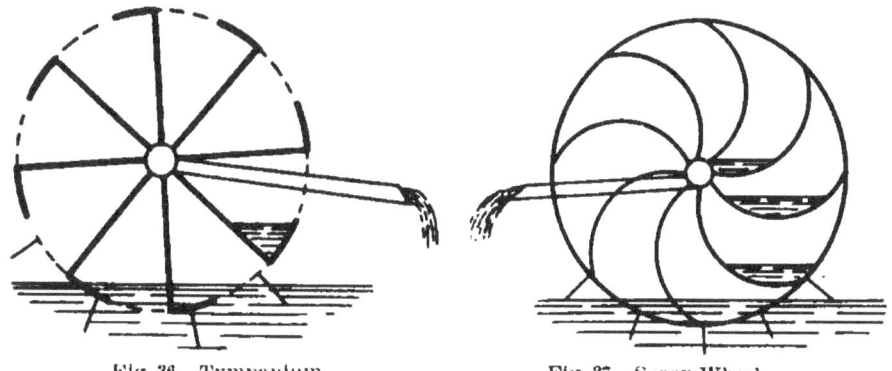

Fig. 36. Tympanium.   Fig. 37. Scoop Wheel.

raised at the extremity of each radius, so that its lever arm, and therefore its resistance, increases as it is raised to a horizontal plane. This defect does not exist in the next type.

71. **Scoop Wheel.** As this wheel (Fig. 37) revolves, the partitions dip into, and scoop up the water; and as they ascend, the water is discharged into a trough placed under one end of the shaft, which is arranged in as many compartments as there are partitions or scoops.

An improved form of scoop wheel is shown in Fig. 38, which consists of four curved scrolls or channels suitably mounted on the wheel body. The water is conveyed to the central chambers by the scrolls, and it then flows away in a channel or trough.

Many other forms of water motor might be shown, most of them ancient and obsolete, which were mainly used for the purpose of raising water; but the above examples serve to indicate some of the principal devices employed for the purpose.

72. Ocean Waves. Many attempts have been made to develop useful power from the almost ceaseless motion of the ocean waves. The essential mechanism usually consists of some form of float which is constrained by a fixed shaft, or a series of such shafts, fastened to a suitable foundation, to move in a vertical direction under the influence of the motion of the waves. The float, by its motion, operates a system of levers and wheels, or ropes and pulleys, which may be made by suitable connections to compress air, or to raise water from a lower to a higher level. In some such way, the irregular or intermittent character of the wave motion may be made to store up power, which, in turn may be released uniformly. Fig. 39 is a diagrammatic representation of such a device.

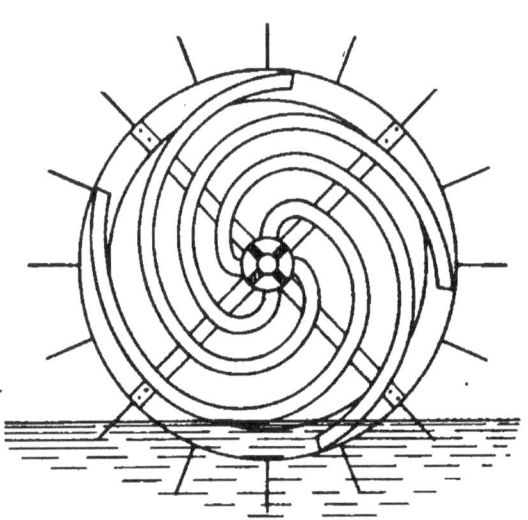

Fig. 38. Scoop Wheel, Improved Type.

73. Tides. The ocean tide furnishes a more reliable means of developing power under suitable conditions. Particularly in the vicinity of tidal rivers, and more rarely along shore, the physical configuration of the land may afford opportunity for impounding large volumes of water during the rising of the tide, which may be made to develop power at ebb by flowing out through a suitable channel and operating one or more wheels. Since the wheels must necessarily remain idle during the rising of the tide, some suitable means must be provided for storing power, so that the machinery dependent upon this power may be in continuous operation, or may operate at any time, irrespective of the tidal conditions. Where power is used intermittently—as in some pumping plants which operate only a certain number of hours each day of 24 hours— a system of power storage, while convenient and advisable to provide against the contingency of a breakdown or other mishap, is not so necessary.

74. Water-Pressure Engine. This is a hydraulic motor which performs work by reason of the static pressure of water acting upon a piston or a revolving disc. The cylinder and piston type of motor has a reciprocating motion identical with that of the steam engine; and the operation is very similar, the water entering and leaving through ports which are opened and closed by valves properly connected with the piston-rod. The useful work is due to the difference in the pressure of admission and discharge. As in the case of the steam engine, the reciprocating motion is generally changed by suitable mechanism into rotary motion before being applied to drive machinery. In the other type, the rotary motion is obtained directly from the shaft of the rotating discs or vanes. This latter type has not been widely used, as in practice there are many inherent difficulties in this mode of transmitting high power.

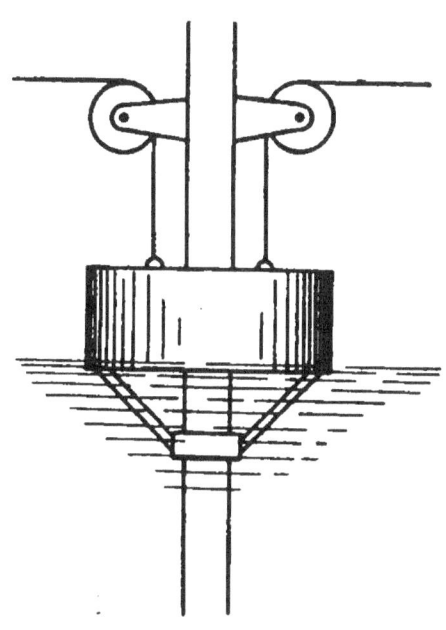

Fig. 39. Device for Utilizing Power of Wave Motion.

IMPULSE WHEELS

75. The term *impulse wheel* is sometimes used to include only those special forms of hydraulic motor which are driven by a jet of water issuing from a nozzle and impinging upon vanes or buckets of special shape attached to the circumference of the wheel. This definition would improperly exclude such motors as the undershot wheel, which is nevertheless a true impulse wheel actuated by a broad stream of water; and also several other types of true impulse wheels.

76. **Horizontal Impulse Wheels.** When a wheel operated by a stream of water issuing from a nozzle and impinging on its vanes is so placed that its plane of rotation is horizontal (the axis being vertical), it is called a *horizontal impulse wheel*.

There are two general classes of such wheels, the *outward-flow*, and the *inward-flow*, as described in Article 52 and illustrated diagrammatically in Figs. 25 and 26. In order to deduce the conditions or relations for maximum efficiency, consider Fig. 26, in which both types are represented, so that the following analysis and the resulting conclusions will be generally applicable to such wheels. The construction of the parallelograms, and the notation, being the same as heretofore, further explanation will be unnecessary.

In order that the water may enter the wheel without shock or foam, the relative velocity V should be tangent to the vane at A as explained before. This condition of tangency will obtain when u and v are proportional to the sines of their opposite angles, in the triangle Auv (as in Article 51, Equations 48 and 48a); that is:

$$\frac{u}{v} = \frac{\sin(\phi - \alpha)}{\sin \phi}; \text{ or, } \cot \phi = \cot \alpha - \frac{u}{v \sin \alpha}$$

The absolute velocity of exit v_1 should be very small (Equations 58 and 59), for the energy represented by this velocity is not given to the wheel, but wasted. Theoretically it should be zero for maximum efficiency, as has already been shown; but practically, if this were the case, the vanes would be unable to clear themselves of the contained water. This absolute velocity v_1 will be small when

$$u_1 = V_1 \quad \ldots \ldots \ldots \ldots (77)$$

These two equations are usually given as representing the conditions of maximum hydraulic efficiency. Equation 77, however, is only approximately true, the real minimum value of v_1 is found when $V_1 = u_1 \cos \beta$, in which case $v_1 = u_1 \sin \beta$; but this equation leads to very complex formulæ. Hence the simpler relation of Equation 77, which is sufficiently accurate, will be used.

Referring to Equation 51, it is clear that if u_1 equals V_1, u must equal V. Then, from the parallelogram at A, Fig. 26, it is seen that when $u = V$, the diagonal bisects the angle ϕ; or,

$$\phi = 2\alpha \quad \ldots \ldots \ldots \ldots (78)$$

Using this value of ϕ in Equation 48, there results:

$$u = \frac{v}{2 \cos \alpha} \quad \ldots \ldots \ldots \ldots (79)$$

Equations 78 and 79 state the conditions involved in Equations 48 and 77, for maximum hydraulic efficiency, in terms sometimes more convenient for use. When a wheel constructed according to this condition (Equation 78) is running with the advantageous velocity u of Equation 79, the absolute velocity of exit is:

$$v_1 = v\frac{r_1}{r}\frac{\sin\frac{1}{2}\beta}{\cos\alpha}; \quad \ldots\ldots\ldots (80)$$

and the corresponding hydraulic efficiency (Equation 59) is:

$$e = 1 - \left(\frac{r_1}{r}\frac{\sin\frac{1}{2}\beta}{\cos\alpha}\right)^2 \quad \ldots\ldots\ldots (81)$$

77. An analysis of this formula teaches that, for high efficiency, both the approach angle α and the exit angle β should be small; but they cannot be zero, otherwise water would not pass into and out of the wheel. Values of 15 to 30 degrees are common. Since, for small angles, the sine varies much more rapidly than the cosine, the equation of efficiency also shows that β is more important than α; so that if β be very small, α may be as large as 40 or 45 degrees, with high efficiency. The equation further shows that for given values of α and β, the inward-flow wheel, in which r_1 is less than r, has a higher efficiency than the outward-flow wheel.

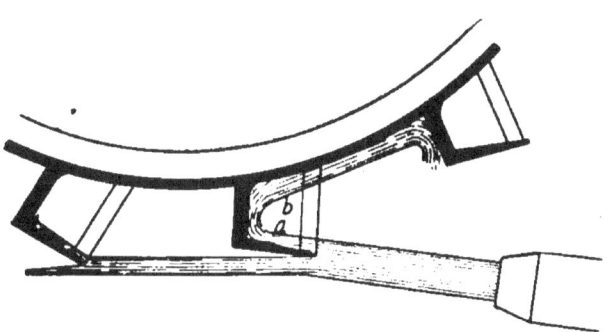

Fig. 40. Faulty Design of Vane.

The actual curve between the entrance and exit points of a vane is not of importance, provided it be smooth and gradual, as abrupt changes of direction lead to shock and to consequent loss of energy.

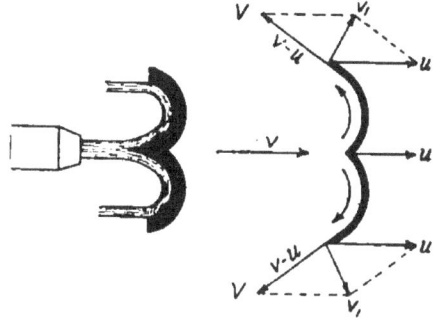

Fig. 41. Good Type of Vane, with Double Cups and Dividing Central Rib.

78. **Vertical Impulse Wheels.** Of this type of wheel (frequently called a *hurdy-gurdy* when the vanes are flat planes, and sometimes a *tangent* or *tangential* impulse wheel), there are several forms in the

market, differing merely in details, and known by various trade names, such as *Pelton, Doble, Cascade*, etc. Essentially this type consists of a wheel mounted on a horizontal shaft, which transmits the power received from a jet or several jets of water acting upon a series of cup-shaped vanes attached to its periphery. The simplest type would be a wheel with flat radial vanes, as in Fig. 21; but, as has already been shown, the efficiency in such a case would be low, so that in practice curved vanes are invariably used.

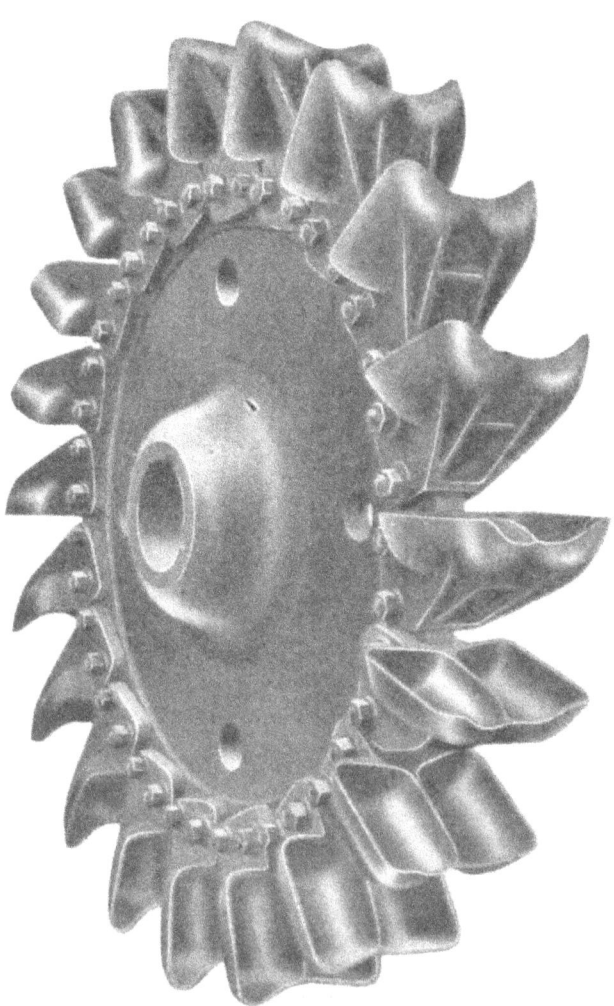

Fig. 42. A 5,000-H.P. Pelton Water-Wheel Runner.
This wheel, 9 ft. 10 in. in diameter, is capable of developing 5,000 h.p. at 325 r.p.m. under 865 feet of effective head.

In Fig. 40 is shown a faulty design of vane, for the water, after striking the outer lip, is abruptly changed in direction at the corners a and b, with consequent shock and loss of energy; also, after leaving a vane, the stream strikes the back of the one adjoining, thus producing back-pressure, with further loss of energy. For these reasons the cups or vanes must be very carefully designed.

In the best forms, the vanes are double cups or buckets with a central rib designed to divide and turn the stream sidewise, while at

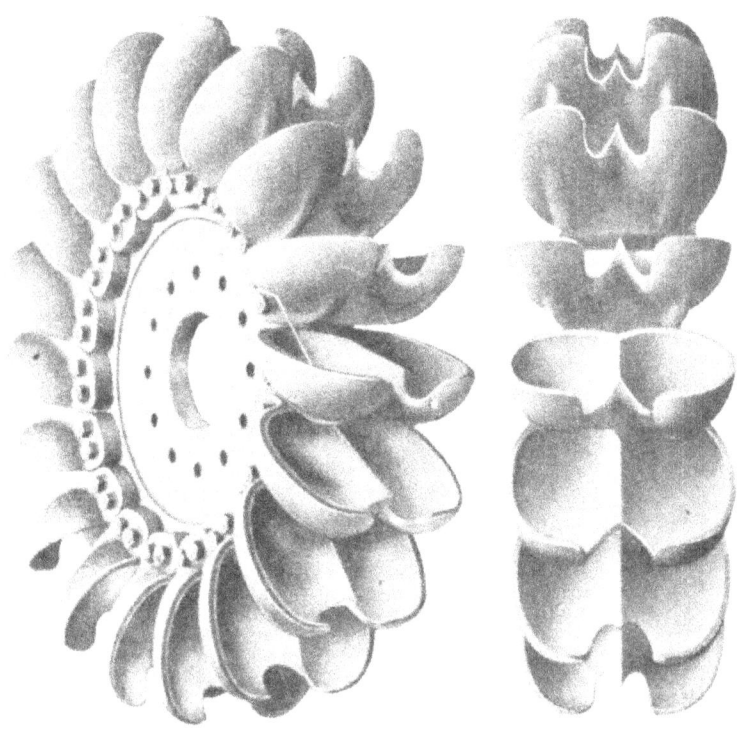

Fig. 43. Runner of 8,000-H.P. Doble Water-Wheel in DeSabla Power Plant.
Velocity of jet, 20,000 ft. per minute.

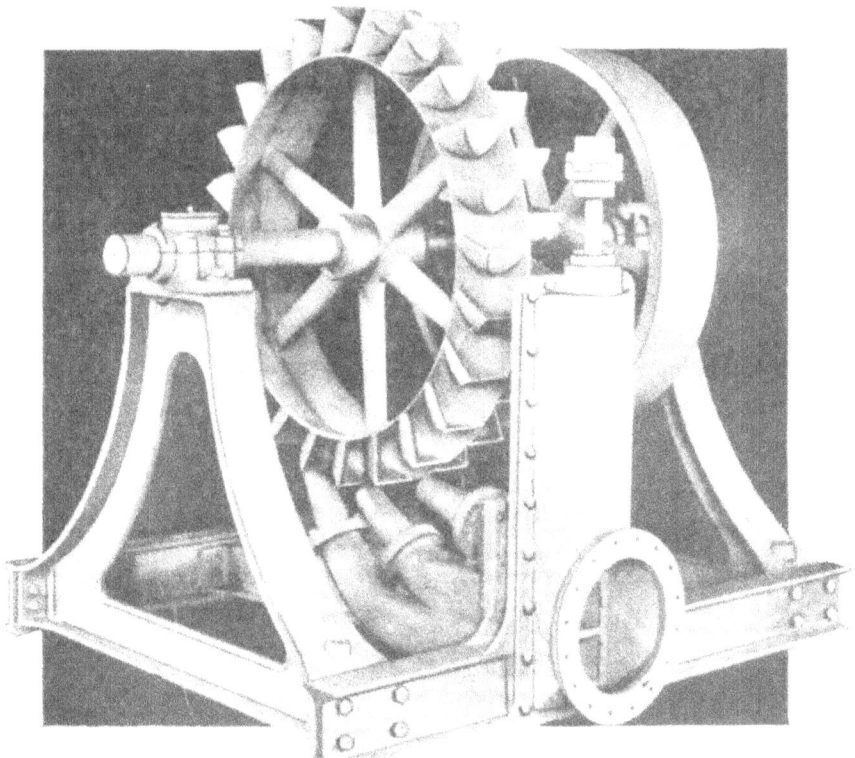

Fig. 44. The "Cascade" (Leffel) Impulse Wheel.
Three-nozzle system.

the same time deflecting it backwards, opposite the direction of motion, as in Fig. 41.

Figs. 42 and 43 show the usual method of attaching the buckets to the wheel; it is clear that in these designs one or more buckets may be very easily and quickly removed and replaced when this is

Fig. 45. A 2,000-Horse-Power Double Unit, 500 Feet Head, for Direct Connection to Generator.

rendered necessary by reason of wear or breakage. The buckets in Fig. 42 are of the *Pelton* type.

In the *Doble* vane, Fig. 43, the outer portion of the lip is dispensed with for the purpose of preventing interference between the jet and the approaching vane, though the central rib is retained for parting the stream sidewise.

In the *Cascade* (*Leffel*) wheel (Fig. 44), the *lobes* or half-buckets are set *staggering*, or *breaking joint*, on opposite sides of a thin circular disc, the sharp edge of which serves the same purpose as the central rib of the other forms in dividing the stream.

79. The analysis and conclusions of Articles 43 and 44, Fig. 23, apply in the case of these wheels; namely, the most advantageous velocity, theoretically, is:

$$u = \tfrac{1}{2}v;$$

and at this velocity, the efficiency is a maximum, and equal to;

$$e \text{ (Max.)} = 1, \text{ or } 100 \text{ per cent,}$$

when $\theta = 180°$—that is, when the stream is completely reversed. However, θ cannot be made equal to 180°, so as to completely reverse the direction of the stream, without interference between the de-

Fig. 46. Interior of Power House of Puget Sound Power Company, Electron, Wash. Four wheel units aggregating 30,000 h.p. in this station; of the "double-overhung" type, coupled to 3,500 k.w. 225 r.p.m. generators. Each unit has an overload capacity of 7,000 h.p.

parting water and the adjoining vane, as shown in Fig. 40, where the water is deflected vertically; and this is equally true when the stream is deflected sideways. The vane is therefore so shaped as to throw the divided stream just clear of the next vane, which condition makes it necessary that θ shall be less than 180 degrees, and consequently the efficiency will be less than 100 per cent, even theoretically. Nevertheless this form of wheel probably comes as near as any to realizing the theoretic condition for maximum efficiency.

As in all the other cases discussed, the theoretic conclusions derived from analyses are not quite true practically. Thus the most advantageous velocity of the wheel is somewhat less than .5 of the jet velocity (though it is probably always considerably greater than .4 that velocity), while the maximum efficiency may be 90 per cent or somewhat higher.

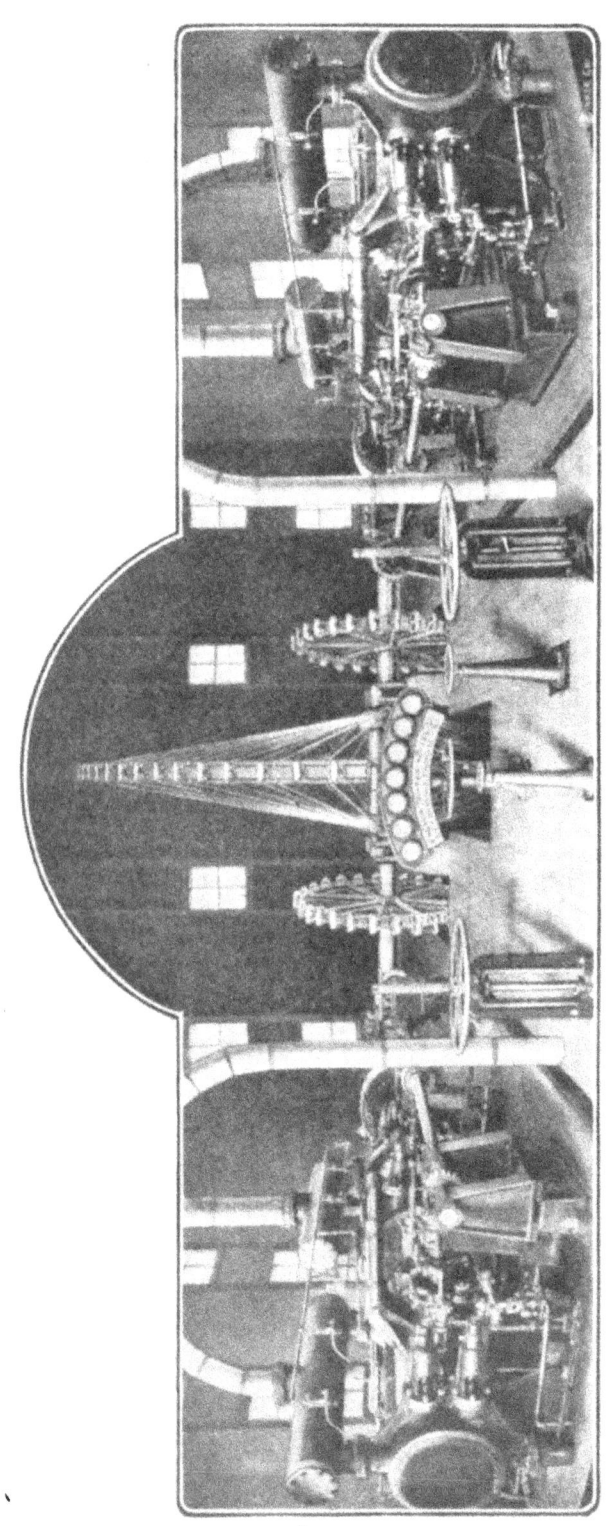

Fig. 47. Hydraulic Air-Compressor Plant. A 1,000-H.P. Duplex Air-Compressor Driven by 3 Pelton Water-Wheels Mounted Direct on Compressor Shaft. Operating under Three Separate Water Heads. Installed at the Morning Mine, Mullan, Idaho.

80. The simplicity, cheapness, and high efficiency of this type of water motor commend it for use when the head of water is not less than about 50 feet—though many are in operation with heads of about 25 feet—especially when the supply of water is not abundant. It has the further advantage, due to its simplicity and cheapness, of allowing of almost indefinite extension of the existing installation, and of division of the power into groups or units, by placing a number of wheels on the same shaft, as in Figs. 45 and 46, or providing a wheel for each machine or group of machines.

Further, several wheels mounted on the same shaft may be operated by jets of water issuing from nozzles under different heads, by properly proportioning the diameters of the wheels and nozzles, as shown in Fig. 47. Here the center wheel is 33 feet in diameter, which is unusually large for this type of motor, and therefore special care was necessary in the design. The two side wheels are each 12 feet in diameter. The variation in heads in this case is about 10 to 1.

For heads much lower than 50 feet, while this type of motor will, with proper regulation, still give a high efficiency, the construction is such that it cannot utilize a large quantity of water, and therefore the power output will not be great. This disadvantage may be obviated to some extent by mounting several wheels on the same shaft; but in the case of low heads, some form of turbine motor is to be preferred.

In setting up, this wheel must of necessity be placed above the tail-race level, and so high above it that there shall be no danger of interference from back-water. This means that a certain proportion of the total available head must be sacrificed to this condition; and unless the total head is sufficiently great to make the loss thus incurred relatively insignificant, this will not be the best type of motor for obtaining the greatest efficiency from the waterfall (see, however, article on "Draft-Tube"). These wheels are well adapted for running high-speed machinery, such as electric generators, air-compressors, etc., by direct connection, thus doing away with much belting or gearing with the attendant loss of power and expense of maintenance. These wheels have been used successfully with heads greater than 2,000 feet. They are manufactured in sizes from 6 inches in diameter to more than 30 feet for special cases, and two or more sizes of nozzle tips are usually provided for adjustment or regulation.

81. *Regulation.* In connection with the practical working of a water-wheel, an important matter is the quick and efficient control of the discharge from the nozzle in order to vary the power output of the wheel as the load varies, or to conform to fluctuations in the supply of water, so as to maintain a constant speed. Interchangeable nozzles of varying sizes have already been referred to; but this method requires hand manipulation, takes time, and requires attention. When the supply of water is adequate, and the power required sufficiently large, or the load variable, from two to five nozzles may be arranged to play simultaneously around the periphery of the wheel, as shown in Fig. 48. By this means, not only may much greater power be derived from one wheel; but, by shutting off one or more jets, the supply and power may be regulated to correspond to the load fluctuations with very little speed variation. Several wheels may be mounted upon the same shaft, each operated by its own jet or jets; and the regulation or control may be effected by shutting off the supply of one or more wheels, which would then run *dead*. In cases where the supply of water is abundant, so that waste is immaterial, good results can be obtained, especially with the smaller wheels, by mounting the two halves of the vanes on separate wheels (practically dividing the ordinary wheel, with its vanes, into two equal portions by a vertical plane at right angles to the axis). When the wheel is working at full power, the two halves are kept together, and thus form an ordinary wheel of this type; when, however, the speed increases, a governor

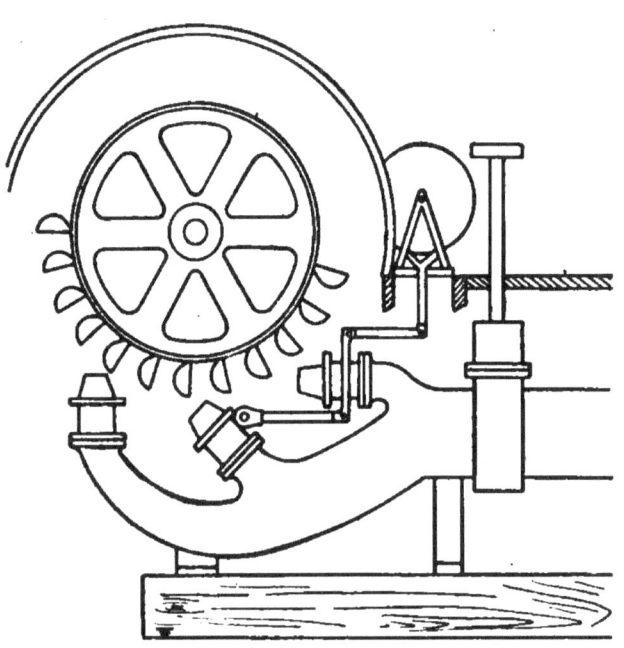

Fig. 48. Wheel Operated by Several Nozzles.

causes the two wheels to separate more or less, and thus some of the water is allowed to escape between. Several other ingenious devices have been developed for the purpose of accomplishing the same end; a description of some of them, taken mainly from manufacturers' catalogues, follows:

82. Under average conditions of operation, a governor is not necessary, as, with a constant load, the speed of the wheel is absolutely uniform. When slight and infrequent changes occur—such as are caused by hanging up stamps of a battery, for example—the wheel can be regulated by hand, by means of the main stop-gate, as shown in Fig. 45; but this would occasion considerable loss of energy, on account of the sudden change of section of the stream. It some-

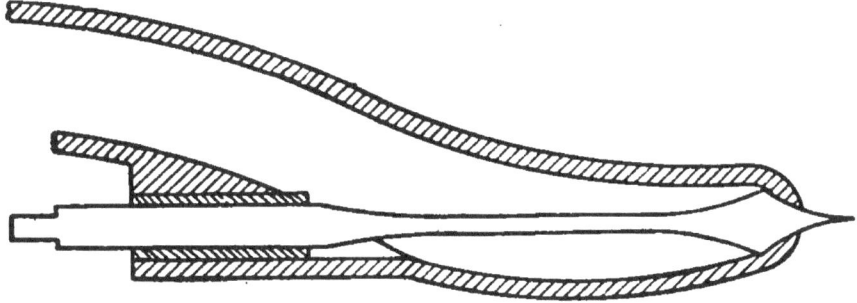

Fig. 49. Section of a Needle Nozzle.

times happens, however, especially when operating electric plants, that the fluctuations in speed are sudden and severe; and in these cases an automatic regulator is essential. In such cases the speed of the wheel may be controlled by means of various devices, among which may be described the following:

The *deflecting nozzle* is a cast-iron nozzle provided with a ball and socket joint, which permits it to be raised or lowered, thus throwing the stream on or off the buckets; the power of the wheel is consequently increased or diminished to correspond to the change of load, and a constant speed is maintained. A steel deflecting plate, which deflects the stream itself—the nozzle remaining stationary—is sometimes used to accomplish the same results when the design will not admit of a deflecting nozzle. Both these devices are wasteful of water; but they effectually prevent *water-hammer*, which would result from a sudden decrease of velocity in the pipe.

The *stream cut-off* is a spherical plate fitting tightly over the end of the nozzle tip, which, by varying its position, changes the discharge area of the nozzle, and thus influences the power of the wheel.

The *needle nozzle* (Figs. 49 and 50) consists of a nozzle body in which is inserted a concentric tapered needle. A change of position of this needle produces a corresponding change of discharge area of the nozzle; the amount of water used is thus varied, and the power of the wheel influenced proportionally.

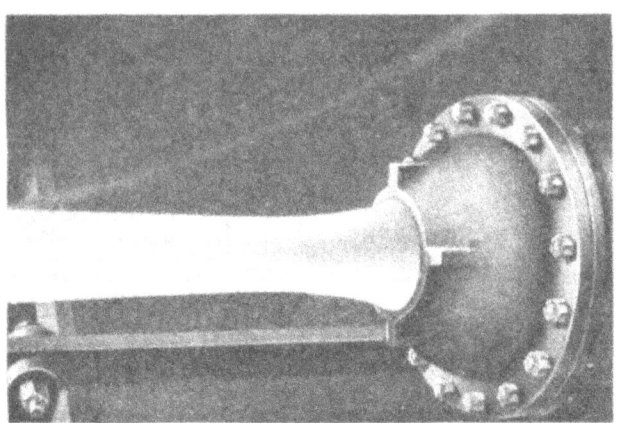

Fig. 50. Stream of Water from Pelton Needle Nozzle Operating under 390-Foot Head and Developing 1,500 H.P.

Note the shadow of needle showing through stream, and the perfect form of jet.

The *needle regulating and deflecting nozzle* (Figs. 51 and 52) is a most valuable combination, consisting of a deflecting nozzle swinging on a pair of trunnions, with which is incorporated a needle nozzle, with means for operating either the needle or deflecting nozzle simul-

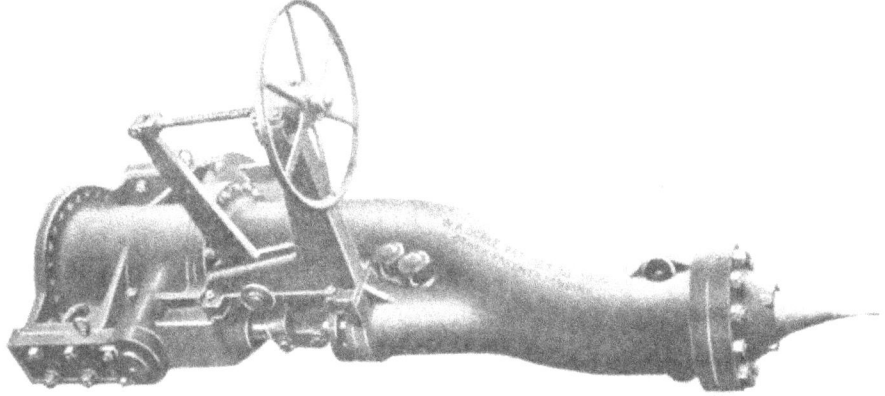

Fig. 51. Doble Needle Regulating and Deflecting Nozzle for 8,000-H.P. Wheel.

taneously or separately. This accomplishes a twofold object—accurate regulation, and water economy without water ram. The deflecting nozzle is a most sensitive means of regulation when actuated by an automatic governor, but does not save water. On the other hand, the needle nozzle, while it is extremely economical in the use of water, is difficult to control quickly by means of the governor. The operation of the combination is as follows:

Assuming the full load to be on the water-wheel, and the nozzle in position of greatest efficiency, a decrease in load, tending to cause increase in speed, will cause the nozzle to be suddenly deflected by the automatic governor.

Fig. 52. Pelton Needle Regulating and Deflecting Nozzle in Operation.

Simultaneously, the needle portion of the nozzle will be actuated by hand, or by another automatic device, tending to close the needle *gradually*

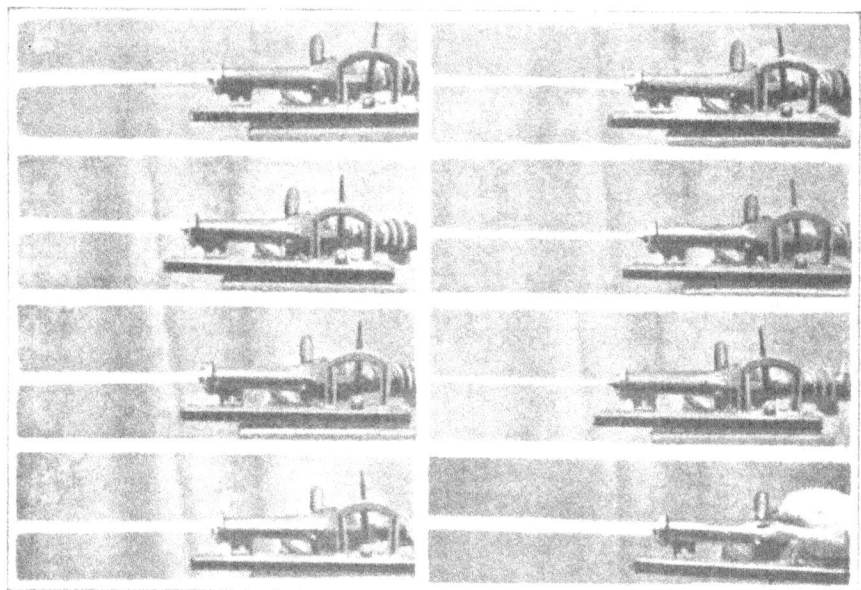

Fig. 53. Various Sized Jets from Small Doble Needle Regulating Nozzle.

and decrease the flow. The governor then raises the nozzle to accommodate the decreased flow of water (and consequent decrease of power), and the nozzle is then brought back to the position of greatest efficiency, having, at the same time, controlled the speed within the required limits.

72   WATER-POWER DEVELOPMENT

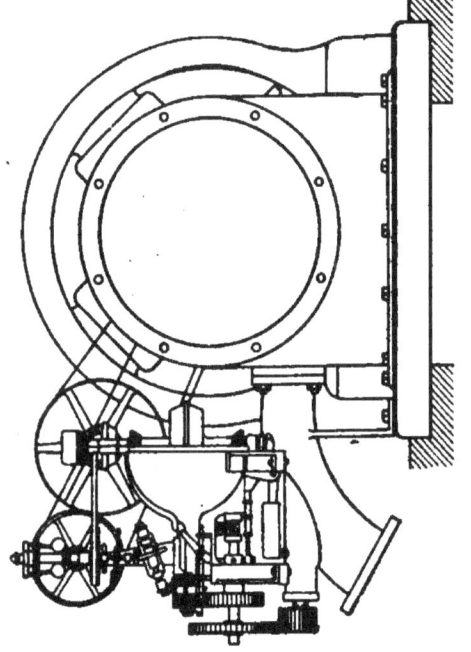
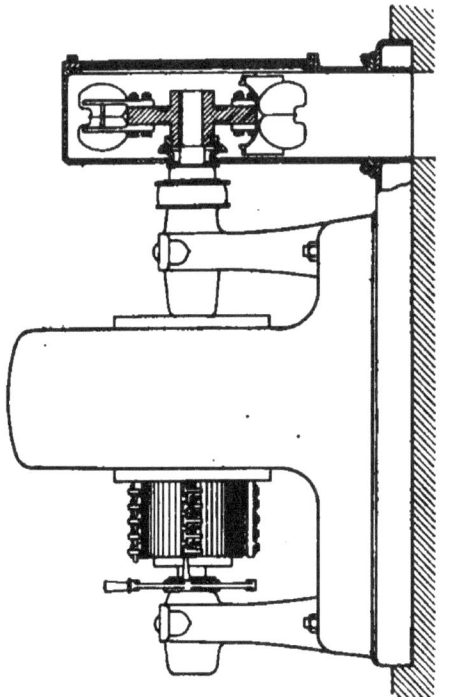

Fig. 54. Hydro-Electric Unit with Needle Nozzle Operated by "Woodward" Governor.

Such a device is essential where water is valuable, and where economy is necessary to carry over the peak load. The needle portion need not necessarily be operated by an automatic device, but may be controlled by hand, and the same results obtained, although necessarily in a longer period of time. In Fig. 52, the upper and lower lines indicate the limits of deflection. Fig. 53 shows how the size of the jet may be varied by means of the needle nozzle.

83. The conditions as to head, power, and character of load determine which device or combination is best suited to any individual case. These various mechanisms are actuated, through a proper system of rock-shafts and levers, by an automatic governor (Figs. 48 and 54), which, for ordinary machinery, may be a *mechanical* governor of the plain, centrifugal-ball type, the power to move the regulating device being furnished directly by the wheel itself; but where close regulation is required, as in driving electrical machinery, a

more sensitive device is necessary. Fig. 55 represents a Lombard automatic governor of the *hydraulic* type, using direct water-pressure to actuate the pistons, which are controlled by balanced valves. Fig. 54 represents a hydro-electric unit in which the needle nozzle, instead of being arranged for hand control, is directly operated by a Woodward compensating governor mounted upon the nozzle body and geared to the needle shaft, which is threaded, and moves in a nut which forms part of the nozzle body, so that the action of the governor regulates directly the position of the needle. It is readily seen that the ball governor is the ultimate device, which actuates or sets in motion the controlling and regulating apparatus. This topic will be further considered under "Turbines" (see Articles 179 *et seq.*).

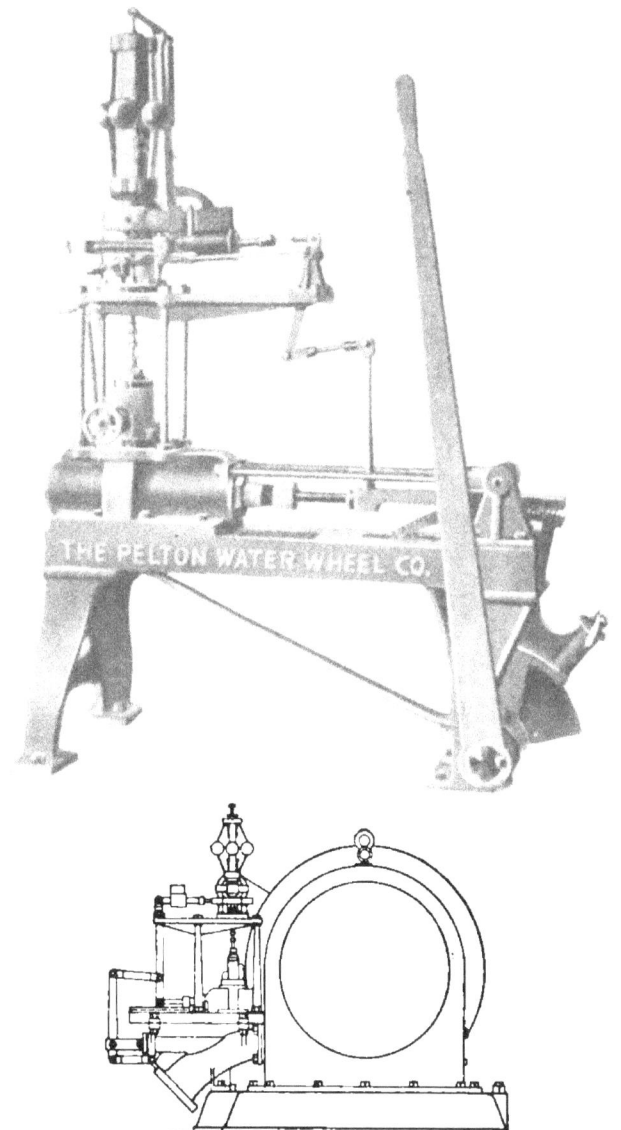

Fig. 55. "Lombard" Automatic Hydraulic Governor.

84. In the case of long pressure-pipes, especially when under high pressure, it is difficult and dangerous suddenly to vary the quantity of water delivered by the nozzle in such a manner as is necessary to regulate the speed of a hydro-electric generating unit subject to

sudden violent variations of load. Consequently it has become customary to regulate the speed of such units by deflecting the jet of water, so that all, or part of it, misses the water-wheel buckets, and is for the moment necessarily wasted. The water which is thus prevented from giving its energy to the water-wheel, is projected through the tail-race at a very high velocity—in some cases exceeding 300 feet per second (18,000 feet per minute)—and becomes destruc-

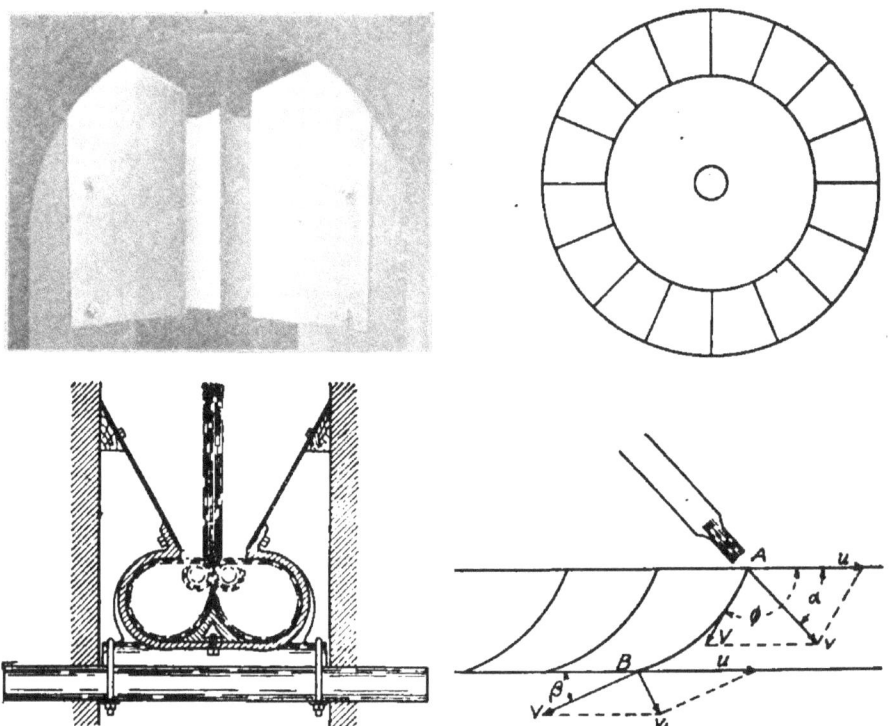

Fig. 56. "Ensign" Vortex Baffle-Plate as Installed in a Tail-Race.

Fig. 57. Downward-Flow Impulse Wheel.

tive, particularly when the water unavoidably carries infinitesimal particles of sand. No masonry can long withstand the action of such a jet, and even iron and steel are rapidly worn away, as if by a terrific sand-blast.

The *Ensign Vortex Baffle-Plate* (patented), illustrated in Fig. 56, is designed to divide such a jet in halves, and deflect the halves until they impinge upon each other, and harmlessly spend their force. The device is a trough-like structure with a sharp central vertical dividing wedge, made to be replaceable in case of wear. The device splits the impinging jet, and guides each half around the curved surfaces, spreading it out into two thin sheets which meet and harmlessly spend their force against each other. The water then falls by gravity into the tail-race with very little disturbance.

85. **Downward-Flow Impulse Wheels.** In this type of motor, the horizontal impulse wheel is driven by the jet from a nozzle inclined downward at a convenient angle, as in Fig. 57, which represents in outline the plan and the development of part of a cylindrical section of such a wheel. The water, in passing through the wheel, neither approaches nor recedes from the axis of rotation; it is therefore sometimes called a *parallel-flow* or *axial-flow* wheel.

The stream enters at A, as shown, with the relative velocity V; passes downward over the vane, always maintaining the same distance from the axis; and, *neglecting the effect of friction and gravity*, issues from the vane at B, with the same relative velocity V.

As before, to prevent impact losses at A, the direction of the relative velocity V must be tangent to the vane at that point; and in order that the efficiency should be high, the absolute velocity of departure v_1 must be small, which later condition will be fulfilled if $u = V$ at B. Therefore, as explained in the preceding analyses, ϕ should be made equal to $2a$, and the best speed of the wheel is $u = \dfrac{v}{2 \cos a}$. The efficiency under these conditions is:

$$e = 1 - \left(\frac{\sin \frac{1}{2} \beta}{\cos \alpha} \right)^2,$$

which again shows that both a and β, particularly the latter, should be small for high efficiency.

In the above analysis, no account was taken of the force of gravity acting as the water descends through the vertical distance between A and B; this would increase the efficiency and the advantageous velocity above the values as found from the equations above.

It is evident that several nozzles might be employed also with this type of wheel, instead of one, where the supply of water is adequate.

(Articles 11 and 12 develop the hydraulic formulæ to be used in problems of nozzle discharge. Article 16 shows the proper relation between the diameters of nozzle and pipe to furnish maximum power; and Article 19 considers the case of multiple nozzles to fulfil the same condition.)

86. Girard Impulse Wheels. This type of wheel (Fig. 58) consists essentially of two flat, parallel, and concentric rings or *crowns*, between which are inserted the curved vanes or blades, the whole attached rigidly to the axle and forming the wheel proper, or *runner*. The feed or operating water issues from a nozzle placed inside the wheel as shown, in which case it is an *outward-flow* impulse wheel; or the nozzle may be placed outside, making it an *inward-flow* wheel; or several nozzles or groups of nozzles may be employed, located symmetrically around the circumference. The analyses and conclusions contained in the preceding articles apply to these cases.

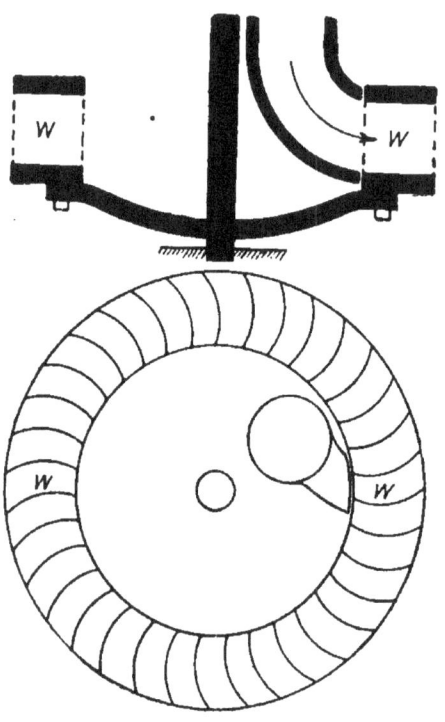

Fig. 58. Girard Impulse Wheel.

Axial or parallel flow may be applied to this type of wheel, as explained in Article 85, under the heading, "Downward-Flow Impulse Wheels."

Occasionally, with the outward-flow type, the two crowns are made to diverge (*i. e.*, their distance apart becomes greater) toward the outer circumference, constituting the so-called *bell-mouthed* profile. In this way, *choking* of the passageways, due to excessive narrowness, is avoided. Openings in the crowns to facilitate the escape of air are frequently made with the same object in view. This type of wheel is widely used in Europe, a number of single motors of this kind developing 1,000 horse-power. Among the several wheels installed at the Terni Steel Works, Italy, ranging from 50 to 1,000 horse-power each, under a head of about 600 feet, is a large 800-horse-power wheel which drives the rolling-mill machinery; its outer diameter is 9 feet 5 inches; its inner diameter, 8 feet 2.4 inches; distance between crowns at entrance, 4.91 inches; at exit, 16.14 inches. The quantity of water used is 16 cubic feet per second; and the normal speed, 200 revolutions per minute.

In the electric power station at Vernayaz, Switzerland, are six 1,000-horse-power Girard wheels, working under a head of 1,640 feet; the outer diameter of each wheel is about 6.5 feet; and the normal speed, about 540 revolutions per minute. These wheels work with but one *guide* (the nozzle tube) each.

A turbine built for the Ouiatchouan Pulp Company (Quebec) has two sets of such guides spaced 180 degrees apart. This wheel develops 1,000 horse-power under a head of 240 feet, running at 225 revolutions per minute; it is enclosed in a cast-iron case and provided with a draft-tube and air-admission valve, both of which contrivances will be described in a later article.

Example 18. Let us assume a Girard outward-flow impulse wheel, with $a = 25$ degrees; $\beta = 20$ degrees; ratio $\dfrac{r_1}{r} = \dfrac{4}{3}$; supplied with 2 cubic feet per second through 12-inch pipe 2,000 feet long, with nozzle attached, having a coefficient of velocity of 0.95. Total head over nozzle tip, 152.00 feet, of which 8.3 feet are consumed in pipe friction and entrance losses. Wheel to make 240 r.p.m.

Velocity in the pipe is:

$$v_p = \frac{q}{a} = \frac{2}{\pi \tfrac{1}{4}} = 2.55 \text{ feet per second.}$$

Velocity of jet is:

$$v = 0.95 \sqrt{2g(152.00 - 8.3) + 2.55^2} = 91.4 \text{ feet per second.}$$

The best speed for the inner rim is

$$u = \frac{v}{2 \cos a} = \frac{91.4}{2 \times 0.906} = 50.5 \text{ feet per second}$$

Since $2\pi r n = 50.5$.

$$r = \frac{50.5}{2\pi \tfrac{240}{60}} = 2.01 \text{ feet.}$$

$$r_1 = \frac{4}{3} \times 2.01 = 2.68 \text{ feet.}$$

The theoretic efficiency is:

$$e = 1 - \left(\frac{r_1}{r} \frac{\sin \tfrac{1}{2}\beta}{\cos a}\right)^2 = 1 - \left(\frac{4}{3} \frac{0.174}{0.906}\right)^2 = 0.93.$$

The actual efficiency would probably have a value between 75 and 80 per cent.

The work imparted to the wheel is, theoretically:

Work $= W \dfrac{v^2}{2g} = 2 \times 62.5 \times \dfrac{\overline{91.4}^2}{64.4} = 16{,}213$ ft.-lbs. per second, if the nozzle be not considered a part of the motor, and if losses in the wheel itself be disregarded.

If the nozzle be considered part of the motor, the work imparted to it, disregarding wheel losses, is:

Work $= wq \left(152.00 - 8.3 + \dfrac{\overline{2.55}^2}{2g} \right)$

$= 62.5 \times 2 \left(143.7 + \dfrac{6.5}{64.4} \right) = 17{,}975$ ft.-lbs. per second.

If the wheel, under the second assumption, have an efficiency of 75 per cent, the useful work of the wheel is:

Useful work $= 17{,}975 \times 0.75 = 13{,}475$ ft.-lbs. per second;

$\dfrac{13{,}475}{550} = 24.5$ horse-power.

87. Rotating Vessels. As a preliminary to the study of the theory and operation of turbines, it will be necessary to consider very briefly some of the essential features of the action of rotating fluids.

Flow from a Revolving Vessel with Free Surface. Let AC (Fig. 59) be any open vessel, revolving about a vertical axis XX. It is shown in treatises on Hydraulics, that the water surface EK, which is horizontal before rotation, becomes, under the action of centrifugal force and gravity, a curved surface AOB, due to the rotation; that this surface is a paraboloid of revolution, so that any vertical section through the axis is a parabola with the vertex at O, and axis vertical and coincident with the axis of rotation; that the head of water over an orifice F in the base or side, at any distance r from the axis of rotation, is $h + \dfrac{u^2}{2g}$, if u be linear velocity of the center of the orifice, and h its distance below

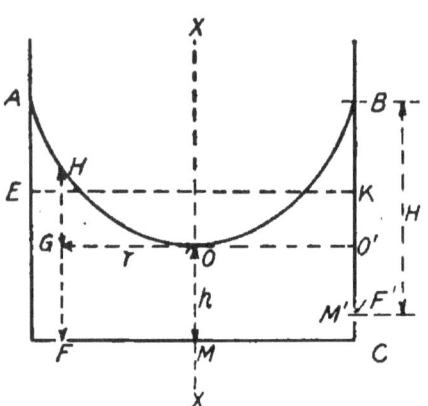

Fig. 59. Revolving Open Vessel.

the lowest point O; and, therefore, that the relative velocity of efflux from F is:

$$V = \sqrt{2g\left(h + \frac{u^2}{2g}\right)} = \sqrt{2gh + u^2} \quad \ldots \quad (82)$$

Let n be the number of revolutions per second; then $u = 2\pi r n$; and

$$V = \sqrt{2gh + 4\pi^2 r^2 n^2} \quad \ldots \ldots \quad (83)$$

This result is independent of the shape of the containing vessel; and the axis of rotation may lie within or without it, the axis of the paraboloid in any case coinciding with the axis of rotation.

88. **Closed Vessel.** The above formulæ apply equally well to the case of a closed rotating vessel in which the curved surface is wholly or partially prevented from forming, as in Fig. 60. Here also h is the depth MO in the axis of rotation; and the parabola AOB represents the vertical section of the paraboloid of *pressures*. In both cases, then,

$$\frac{u^2}{2g} = \frac{2\pi^2 r^2 n^2}{g}$$

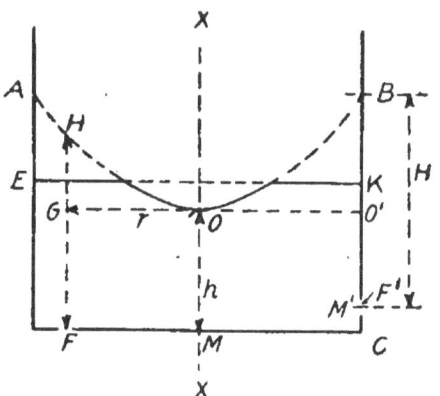

Fig. 60. Revolving Closed Vessel.

is the head GH, to be added to the minimum static head MO at the axis, to obtain the total pressure head over the orifice. If the orifice is in the vertical bounding wall of the vessel, as at F^1, the pressure head is $M^1 O^1 + O^1 B = H$.

89. **Revolving Tubes.** Fig. 61 represents the simple case of one or more hollow arms attached to a vessel, and rotating with it about a vertical axis. From what has preceded, it is clear that the static pressures at the points A and B in the tube, when rotation has been established, but *when no flow occurs*, are, respectively:

$$OM + GH = h + \frac{u^2}{2g}, \text{ for the point } A; \text{ and,}$$

$$OM + G_1 H_1 = h + \frac{u_1^2}{2g} \text{ for the point } B,$$

if u and u_1 be the linear velocities of the points A and B respectively.

When the orifices are opened and *flow takes place*, the pressure-head in each case falls by an amount equal to the velocity head *plus* the head lost in frictional resistances, as explained in Articles 6 and 7; and the line of pressure now assumes some other form, such as LB. Neglecting for the present the frictional losses, it is evident that the following relations must obtain, by reason of the principle of the conservation of energy:

$$h + \frac{u^2}{2g} = h' + \frac{V^2}{2g},$$

which becomes, for the point B, since no pressure head exists at the end of the tube when it discharges freely into the air:

$$h + \frac{u_1^2}{2g} = 0 + \frac{V_1^2}{2g}, \quad \ldots\ldots\ldots\ldots (83a)$$

so that

$$h = h' + \frac{V^2}{2g} - \frac{u^2}{2g} = \frac{V_1^2}{2g} - \frac{u_1^2}{2g}. \ldots (83b)$$

if V and V_1 represent the relative velocities in the tube at the points A and B respectively. If the tube is submerged as in Fig. 76, there is static pressure at the end; so that, if h'' is the static pressure on the end (the depth of submergence), then,

$$h + \frac{u_1^2}{2g} = h'' + \frac{V_1^2}{2g}, \quad \ldots\ldots\ldots\ldots (84)$$

and therefore,

$$h = h' + \frac{V^2}{2g} - \frac{u^2}{2g} = h'' + \frac{V_1^2}{2g} - \frac{u_1^2}{2g}. (84a)$$

The above equation (84a) expresses the relation between the pressure-head, velocity-head, and rotation-head at any point of a revolving tube. In case the tube is only partly full, as when a stream impinges and glides along a vane (or one side of a tube or bucket of a water-motor), there can be no static pressure, and the above becomes:

$$V_1^2 - V^2 = u_1^2 - u^2, \quad \ldots\ldots\ldots\ldots (84b)$$

which is Equation 51, for the case of a jet impinging on a vane.

Fig. 61 represents essentially a reaction wheel, since the dynamic pressure causing rotation is caused entirely by the reaction of the issuing jets.

90. In order to discuss the work and energy of such an apparatus, we may use Equation 57, which expresses the work of the impulse of the entering stream and the reaction of the departing stream, by simply omitting the term representing the former. Accordingly, for the work of a reaction wheel:

$$\text{Work} = W\frac{-u_1 V_1 \cos \theta}{g} \quad \ldots \ldots \ldots \ldots \ldots \ldots (85)$$

$$= W\frac{u_1 V_1 \cos \beta - u_1^2}{g} \text{(from Equation 55)} \ldots \ldots (86)$$

$$= W\frac{u_1 \cos \beta \sqrt{2gh + u_1^2} - u_1^2}{g} \text{(from Equation 83a)} \cdot (87)$$

Dividing the expression for Work by the theoretic energy Wh, we have:

$$\text{Efficiency} = \frac{u_1 \cos \beta \sqrt{2gh + u_1^2} - u_1^2}{gh} \ldots \ldots \ldots \ldots (88)$$

The work is zero when $u_1 = 0$—that is, when there is no rotation; also when $u_1^2 = 2gh \cot^2 \beta$; and it is a maximum, and equal to

$$\text{Work (Max.)} = Wh(1 - \sin \beta) \cdot \ldots \ldots \ldots \ldots \ldots \ldots (89)$$

when,

$$u_1^2 = \frac{gh}{\sin \beta} - gh, \ldots (90)$$

the efficiency, in this case, being:

$$e \text{ (Max.)} = 1 - \sin \beta \quad \ldots (91)$$

The work and efficiency, therefore, increase as the angle β decreases. When $\beta = 90$ degrees, the work and the efficiency both become zero, for the jet in such case issues radially; when $\beta = 0$ degrees, the work is Wh, and the efficiency is unity, or 100 per cent; but the velocity u_1 (*and therefore also* V_1) becomes infinitely great. It must be remembered that frictional and air resistances have not been considered in the above analysis; both increase rapidly with increased speed of rotation. In general, however, it may be stated that within certain limits the efficiency of a reaction wheel increases with the speed and with the smallness of the angle β; and it is greatest in any given case, when the angle β is zero—that is, when the water issues in a direction exactly opposite to that of rotation.

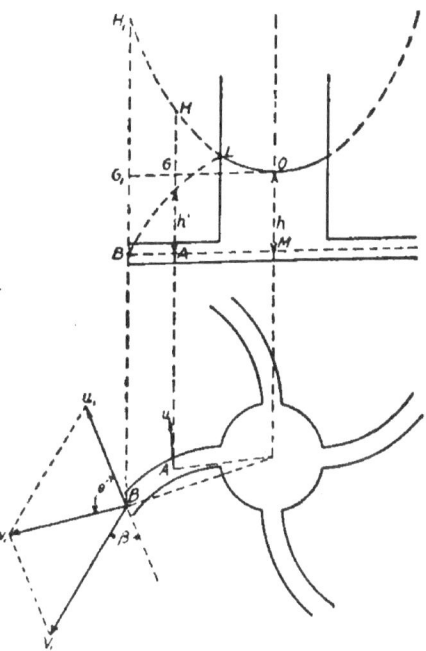

Fig. 61. Revolving Vessel with Hollow Arms Attached.

91. Reaction Wheel. Fig. 62 represents an apparatus commonly known as *Barker's Mill*. It is the reaction wheel described in the preceding article, with the direction of the issuing streams of water directly opposite to that of revolution, or $\beta = 0$. Making $\beta = 0$ in the preceding equations, we have:

$$\text{Work} = W \frac{u_1 \sqrt{2gh + u_1^2} - u_1^2}{g}; \quad \ldots \ldots (92)$$

$$\text{Efficiency} = \frac{u_1 \sqrt{2gh + u_1^2} - u_1^2}{gh}; \quad \ldots \ldots (93)$$

$$\text{Work (Max.)} = Wh; \quad \ldots \ldots \ldots \ldots \ldots \ldots (94)$$

$$\text{Efficiency (Max.)} = \text{unity, or 100 per cent}, \ldots (95)$$

when $u_1 = $ infinity; in which case also $v_1 = $ infinity.

If a_1 be the area of the exit orifices, and w the weight of a cubic unit of water, the weight of water discharged in one second is $w a_1 v_1$,

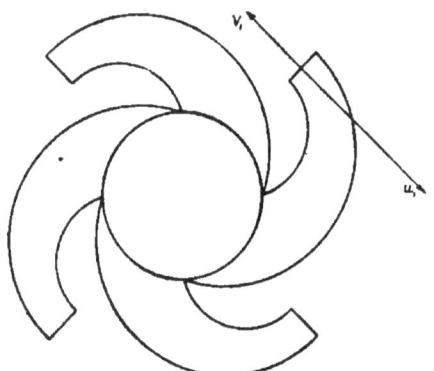

Fig. 62. Barker's Mill.

which becomes infinite when $u_1 = V_1 = $ infinity. As stated before, frictional and air resistances increase rapidly with the speed, so that the above relations, in deriving which these resistances have not been considered, are theoretic. It is evident, however, that the efficiency of a reaction wheel of this type increases with the speed within certain limits; and that the discharge varies with the speed.

92. **Effect of Friction.** If c_v be the coefficient of velocity representing the effect of friction in the arms and orifice, then,

$$V_1 = c_v \sqrt{2gh + u_1^2}, \quad \ldots \ldots \ldots (96)$$

instead of the theoretical expression,

$$V_1 = \sqrt{2gh + u_1^2}$$

The expressions for the effective work of the wheel and the efficiency then become:

$$\text{Work} = W \frac{c_v u_1 \sqrt{2gh + u_1^2} - u_1^2}{g} \ldots (97)$$

$$\text{Efficiency} = \frac{c_v u_1 \sqrt{2gh + u_1^2} - u_1^2}{gh}, \quad \ldots (98)$$

$$\text{Efficiency (Max.)} = 1 - \sqrt{1 - c_v^2} \ldots \ldots \ldots (99)$$

when,

$$u_1^2 = \frac{gh}{\sqrt{1-c_v^2}} - gh \cdots \cdots (100)$$

If $c_v = 1$—that is, when frictional loss is not considered—$e = 1$; and $u_1 = V_1 =$ infinity, as before. When $c_v = .94$, the advantageous velocity $u_1 = \sqrt{2gh}$, and the efficiency is 65 per cent. Thus the effect of friction is greatly to decrease the theoretic efficiency. To render c_v large, the tubes should be smooth and well rounded by means of easy curves. In addition to the above considerations, the air resistance, which has not been included in the above analysis, increases very rapidly with the speed of rotation,

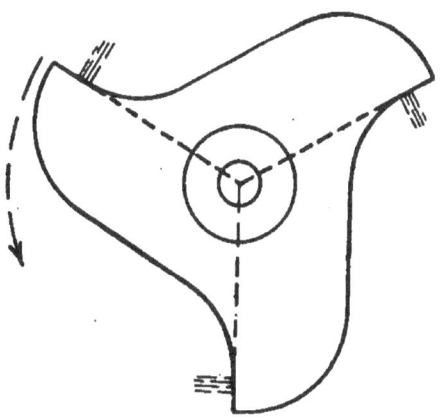

Fig. 63. Scotch Wheel.

and its effect is to reduce still further the computed efficiency. Because of the low actual efficiency resulting from the above factors, the reaction wheel is not used as a practical hydraulic motor.

93. The *Scotch wheel* (Fig. 63) is an improvement on the Barker's Mill; the three orifices are made adjustable in size by means of movable flaps, for the purpose of regulating the quantity of water and the power.

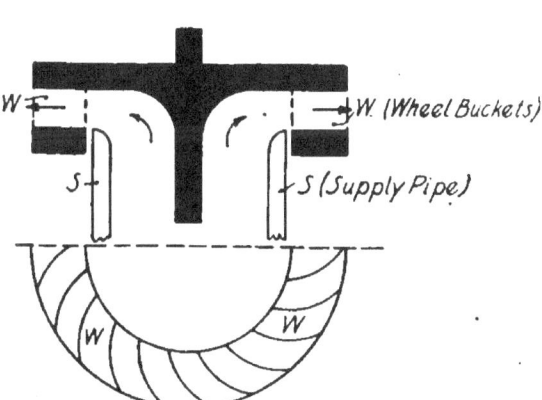

Fig. 64. Combe's Turbine.

94. The next advance in hydraulic motor design consisted in the employment of a large number of issuing streams, as in the *Combe's turbine* (Fig. 64). Here the supply pipe furnishes water directly to the buckets, without the directing intermediary of guides; the water completely fills the passageways, and discharges into the atmosphere with a somewhat low absolute velocity. A modification of this type, with supply pipe above, was called the *Cadiat turbine*.

95. "In 1826 the French engineer Fourneyron improved the Cadiat turbine by placing fixed guide-blades just inside the wheel-ring, around the entire circumference, by means of which the water received a forward direction of motion before entering the channels of the moving turbine. This rendered attainable a very low value of the absolute velocity of the water at exit from the outer rim of the wheel-ring. Also, the wheel being operated under water, the complete filling of the wheel-channels was assured when properly designed. This was the first modern turbine—a motor which, as varied and improved by Fontaine, Henschel, Jonval, and others in Europe, and by Boyden and Francis, and their successors in America, has grown in popular favor, and, together with the impulse wheels already described, has almost entirely supplanted the old forms of vertical water-wheels so long considered as giving the highest efficiency."*

*Church, "Hydraulic Motors."

EXAMINATION PAPER

WATER-POWER DEVELOPMENT

PART I

Read carefully: Place your name and full address at the head of the paper. Any cheap, light paper like the sample previously sent you may be used. Do not crowd your work, but **arrange it** neatly and legibly. *Do not copy the answers from the Instruction Paper; use your own words, so that we may be sure that you understand the subject.*

1. Define the terms *Work, Power, Energy*. What are their units of measurement?

2. Illustrate the relation between *Pressure-Head, Velocity Head*, and *Gravity-Head*.

3. Define, and explain the meaning of the term *Efficiency* of a hydraulic motor. Why is it that a motor can never work with an efficiency of 100 per cent?

4. What is the object of attaching a nozzle to the end of a pipe?

5. Suppose, in Example 8 (p. 11), the nozzle to be removed from the end of the pipe. Find the velocity of flow, and the discharge.

6. What should determine the proper relation between the pipe and nozzle diameters in any given case?

7. What is the proper relation between total head and friction head to furnish maximum power in the case of a pipe with and without a nozzle attachment?

8. When multiple nozzles are employed, of the same size or of different sizes, what principle is used to determine their proper diameters to furnish maximum power?

9. An iron pipe is 1,200 feet long and 1 foot in diameter, and its extremity is under a static head of 100 feet. What size nozzle is required for maximum power? What is the velocity of the jet, and in the pipe? What is the discharge? If there are to be two nozzles of the same size, what are their diameters? Three nozzles of the same size? If there are to be two nozzles, and one is required to be 4 inches in diameter, what must be the diameter of the other?

10. Distinguish between the term *Impulse* and *Reaction*; *Static* and *Dynamic Pressure*. To give an impulse of 120 lbs., what must be the velocity of a jet of water issuing from a nozzle 1.25 inches in diameter?

11. What is the impulse of the jet of water in Example 7 (p. 9)? What is the back-pressure on the pipe exerted by the jet?

12. Explain (with illustrations) how the dynamic pressure exerted by a jet of water may be measured.

13. Suppose the jet of water of Example 8 (p. 11) impinges on the surface shown in Fig. 7 (p. 19), what would be the value of the weight P to maintain equilibrium? What would be the value in a case like that of Fig. 9 (p. 20)?

14. Define the terms *Absolute Velocity* and *Relative Velocity*.

15. Give the reasons why the dynamic pressure exerted by a jet of water on a stationary surface is greater than on a similar surface in motion away from the jet.

16. Define the term *Hydraulic Motor*; and state what, in general, is the difference between a water wheel and a turbine.

17. Discuss the requirements for *high efficiency* in hydraulic motors, explaining the causes of energy losses.

18. Compare the relative advantages and disadvantages of *overshot*, *breast*, and *undershot* wheels.

19. In the case of the overshot-wheel, analysis shows certain losses to be inevitable. What are they? And how are they reduced to the smallest possible limits?

20. What conditions should be observed in installing overshot, breast, and undershot wheels, respectively? What actual efficiencies may be expected in each of the above types of wheel?

21. Describe the different forms of bucket used in impulse wheels.

22. What are the essential conditions to be observed in the design of an efficient bucket for an impulse wheel?

23. What are the peculiar advantages of the vertical type of impulse wheel?

24. Describe the various methods of governing and regulating in the case of vertical impulse wheels, giving their advantages and disadvantages.

25. Describe briefly a downward-flow impulse wheel.

26. Describe briefly a Girard impulse wheel.

27. Fig. 60 represents a closed vessel completely filled with water, rotating around its central vertical axis with a speed of 100 revolutions per minute. The orifice F' is 2 feet below the upper surface, and 3 feet from the axis of rotation. Compute the theoretic absolute and relative initial velocity of efflux. What would be their values if the vessel were stationary? If the coefficient of velocity for the orifice were 0.96, what would be the actual velocities in the two cases?

28. Using the appropriate equations for the purpose, discuss the topic *Work and Efficiency* in the case of a reaction wheel, both with and without a consideration of frictional resistances.

29. Explain why the reaction wheel is not used as a practical hydraulic motor.

30. Describe the transition types between the Barker's Mill and Fourneyron's first turbine, including the latter.

After completing the work, add and sign the following statement:
I hereby certify that the above work is entirely my own.

(Signed)

DIVERTING DAM, INTAKE, AND HEAD-GATE FOR FLUME
Courtesy of Pelton Water Wheel Co., San Francisco, Cal.

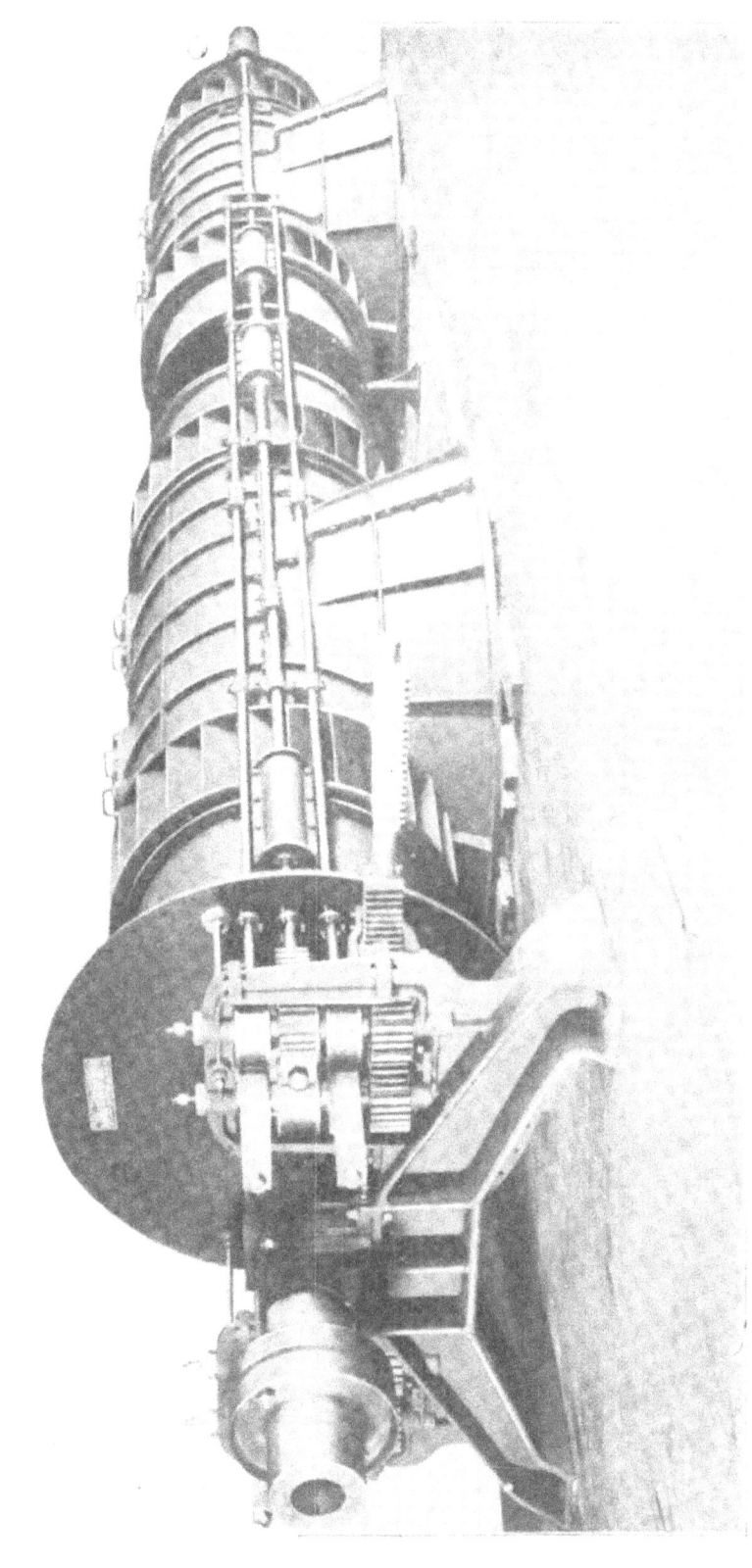

UNIT CONSISTING OF TWO PAIRS OF 51-INCH "CYLINDER GATE NEW AMERICAN" TURBINES ON HORIZONTAL SHAFT
Designed for direct connection to 1,500-K. W. generator, operating under a variable head of 25 to 35 feet.
Courtesy of Dayton Globe Iron Works Company, Dayton, Ohio.

www.ingramcontent.com/pod-product-compliance
Lightning Source LLC
Chambersburg PA
CBHW082347220526
45470CB00008B/2674